THE FRACTALIST

THE
FRACTALIST

Memoir of a Scientific Maverick

Benoit B. Mandelbrot

PANTHEON BOOKS

New York

All rights reserved. Published in the United States by Pantheon Books,
a division of Random House, Inc., New York, and in Canada
by Random House of Canada Limited, Toronto.

Pantheon Books and colophon are registered trademarks
of Random House, Inc.

Library of Congress Cataloging-in-Publication Data
Mandelbrot, Benoit B.
The fractalist : memoir of a scientific maverick /
Benoit Mandelbrot.
p. cm.
Includes index.
ISBN 978-0-307-37735-7
1. Mandelbrot, Benoit B.
2. Mathematicians—France—Biography.
3. Fractals. I. Title.
QA29.M34A3 2012
510.92—dc22
[B]   2012017896

www.pantheonbooks.com

Front-of-jacket image Benoit Mandelbrot. Emilio Segre Visual Archives/
American Institute of Physics/Photo Researchers, Inc.
Jacket design by Peter Mendelsund

Printed in the United States of America
First Edition
2   4   6   8   9   7   5   3   1

*My long, meandering ride through life
has been lonely and often very rough.
Without loving help, it would have been short,
nasty, and unproductive. But I have been lucky.
Father and Mother taught me the art of survival.
Uncle took me as an unruly but grateful student.
Aliette later joined them, and she, our sons, and
our grandchildren taught me how to smile.
To my steady beacons, these scenes
from a life are dedicated.*

*This is a memoir of an ardent but bumpy
pursuit of order and beauty in roughness—
through mathematics and economics, the
sciences, engineering, and the arts.
It led me to encounter more than my share of
unusually diverse and forceful persons along the way.
Many were warm and welcoming; many were
indifferent, dismissive, hostile—even beastly.
This book cannot possibly mention them all,
but every one taught me something
and to all I owe a great deal.*

*To the memory of Johannes Kepler,
who brought ancient data and ancient
toys together and founded science.*

Contents

CONTENTS

Part Three: My Life's Fruitful Third Stage

Acknowledgments

by Aliette Mandelbrot

MY HUSBAND DIED shortly before *The Fractalist* went to the publisher. Benoit spent years writing this memoir. It was a labor of love for him, and my deepest gratitude goes out to the people who kept his words alive.

Without Merry Morse, Benoit's able assistant, this memoir would never have been finished. Merry put her astonishing editorial and problem-solving skills to work, preserving his words, style, and spirit. Thank you, Merry, for your hard work, hours beyond counting, coordination with the publishers, and friendship.

Michael Frame, Benoit's close friend and colleague at Yale, penned the afterword, consulted with us on the fractal images and technical sections throughout the book, and was always there with ready answers to our seemingly endless list of questions.

Richard Voss, a colleague and friend at IBM Research in Yorktown Heights, reviewed and clarified the section on early computer graphics at IBM, and explained the technology behind the images he generated.

Alan Norton, another colleague of Benoit's at IBM Research, reviewed sections of the book related to Julia sets and suggested ideas for using related images.

Benoit's collaborator on *The (Mis)Behavior of Markets,* Richard Hudson, reviewed the sections on finance.

Lisa Margolin searched many hours to find a 1938 map of Byelorussia showing Połoczanka, where Benoit spent one summer of his childhood.

Irene Greif, director of IBM Research in Cambridge, and Charles Lickel, former vice president at IBM Research in Hawthorne, generously gave time to Merry Morse to work on this book.

A C K N O W L E D G M E N T S

Maida Eisenberg welcomed Benoit to IBM Research in Cambridge, offering him a congenial and comfortable setting to work on this memoir.

Paul Moody, a research designer at IBM Research in Cambridge, prepared some of the fractal images for this book.

Jane Olingy, administrator at IBM Research in Cambridge, prepared and sent all drafts to the publisher.

Special thanks to Dan Frank at Pantheon, for his invaluable help in shaping the book.

And finally, we appreciate the Alfred P. Sloan Foundation, whose grant supported this book.

Beauty and Roughness

Introduction

NEARLY ALL COMMON PATTERNS IN NATURE are rough. They have aspects that are exquisitely irregular and fragmented—not merely more elaborate than the marvelous ancient geometry of Euclid but of massively greater complexity. For centuries, the very idea of measuring roughness was an idle dream. This is one of the dreams to which I have devoted my entire scientific life.

Let me introduce myself. A scientific warrior of sorts, and an old man now, I have written a great deal but never acquired a predictable audience. So, in this memoir, please allow me to tell you who I think I am and how I came to labor for so many years on the first-ever theory of roughness and was rewarded by watching it transform itself into an aspect of a theory of beauty.

* * *

The broad-minded mathematician Henri Poincaré (1854–1912) remarked that some questions one chooses to ask, while others are "natural" and ask themselves. My life has been filled with such questions: What shape is a mountain, a coastline, a river, or a dividing line between two river watersheds? What shape is a cloud, a flame, or a welding? How dense is the distribution of galaxies in the universe? How can one describe—to be able to act upon—the volatility of prices quoted in financial markets? How to compare and measure the vocabularies of different writers? Numbers measure area and length. Could some other number measure the "overall roughness" of rusted iron, or of broken stone, metal, or glass? Or the complexity of a piece of music or of abstract art? Can geometry deliver what the Greek root of its name seemed to promise—truthful measurement, not only of cultivated fields along the Nile River but also of untamed Earth?

These questions, as well as a host of others, are scattered across a multitude of sciences and have been faced only recently . . . by me. As an adolescent during World War II, I came to worship a major achievement of a mathematician and astronomer of long ago, Johannes Kepler (1571–1630). Kepler combined the ellipses of ancient Greek geometers with a failure of ancient Greek astronomers, who mistakenly believed that persistent "anomalies" existed in the motion of planets. Kepler used his knowledge of two different fields—mathematics and astronomy—to calculate that this motion of the planets was not an anomaly. It was, in fact, an elliptical orbit. To discover something like this became my childhood dream.

A most impractical prospect! Not one leading to a career in any organized profession, nor providing a way of shining in life—a prospect that my uncle Szolem, an eminent mathematician, repeatedly called completely childish. Yet somehow fate did allow me to spend my life pursuing that dream. Through extraordinarily good fortune, and a long and achingly complicated professional life, it was eventually fulfilled.

In my Keplerian quest I faced many challenges. The good news is that I succeeded. The bad news, or perhaps additional good news, is that my "success" raised a host of new and different problems. Moreover, my contributions to seemingly unrelated fields were actually closely related and eventually led to a theory of roughness—a challenge dating back to ancient times. The Greek philosopher Plato had outlined this challenge millennia before our time, but nobody knew how to pursue it. Was I that person?

* * *

An acquaintance of mine was a forceful dean at a major university. One day, as our paths crossed in a busy corridor, he stopped to make a comment I never forgot: "You are doing very well, yet you are taking a lonely and hard path. You keep running from field to field, leading an unpredictable life, never settling down to enjoy what you have accomplished. A rolling stone gathers no moss, and—behind your back—people call you completely crazy. But I don't think you are crazy at all, and you must continue what you are doing. For a thinking person, the most serious mental illness is not being sure of who you are. This is a problem you do not suffer from. You never need to reinvent yourself to fit changes in circumstances; you just move on. In that respect, you are the sanest person among us."

Quietly, I responded that I was not running from field to field, but rather working on a theory of roughness. I was not a man with a big hammer to whom every problem looked like a nail. Were his words meant to compliment or merely to reassure? I soon found out: he was promoting me for a major award.

Is mental health compatible with being possessed by barely contained restlessness? In Dante's *Divine Comedy*, the deceased sentenced to eternal searching are pushed to the deepest level of the Inferno. But for me, an eternal search across countless scientific fields beyond obvious connection managed to add up to a happy life. A rolling stone perhaps, but not an unresponsive one. Overactive and self-motivated, I loved to roll along, stopping to listen and preach in lay monasteries of all kinds—some splendid and proud, others forsaken and out of the way.

* * *

At age twenty, I was one of twenty men who won entry into the most exclusive university in France, the École Normale Supérieure. When I retired at eighty, I was in the mathematics department at Yale as Sterling Professor—one of about twenty people at Yale's highest rank. I entered and left "active life" under the most exclusive and noncontroversial conditions possible. And along the way I did gather some "moss."

My life since age thirty-five—a turning point—has been most atypical in different but fruitful ways. It reminds me of that fairy tale in which the hero sees a small thread where none was expected, pulls on it, harder and harder, and unravels a variety of wonders beyond belief . . . all totally unexpected. Examined one by one, these wonders of mine "belonged" to fields of knowledge far removed from one another. One could pursue each on its own, to great benefit, as I did early on in my career. But I later adopted a broader point of view, for which I was well rewarded. All those contributors to different fields were easiest to study when recognized as "peas in a pod," pearls of all sizes from a very long necklace.

Do those fields seem far removed from one another? Did I scatter my efforts to self-destructive excess? Possibly. Tight and deliberate self-control kept me focused on those rough shapes that had no common name but begged for one. Bringing these separate fields together put me, step by step, in the unexpected, rare, and dangerously exposed position of opening a new field and gaining the right to name it. I called it fractal geometry.

Every key facet of fractal geometry suffers from a quandary that physicists of the early 1900s called a "catastrophe." The theories of that time predicted an infinitely large value for energy radiated by certain objects. In reality, this was not the case, so something had to give! Solving this quandary was achieved by quantum mechanics, one of the major revolutions of twentieth-century physics and the foundation of much of modern technology, including computers, lasers, and satellites.

What unified all my "peas" was the opposite end of the same quandary. Many domains of science that I dealt with centered around quantities that were assumed to have well-defined finite values, such as lengths of coastlines. However, those finite values resisted being pinned down. Measuring the length of a coastline with shorter measuring rods detects smaller features, leading to longer measurements. The insight that let me study those fields was that one should allow those key quantities to be infinite.

* * *

How did this all come to be? Uncle Szolem and I were both born in Warsaw. We each had a good eye and became counted as mathematicians. But the overly interesting times that cursed his teens and later mine, helped shape us into altogether different people. He found fulfillment as a sharply focused establishment insider, while I thrived as a hard-to-pigeonhole maverick.

As an adolescent during World War I, Uncle roamed around a Russia in the throes of revolution and civil war. He was introduced early to a well-defined and nonvisual topic: classical French mathematical analysis. He fell in passionate lifelong love with it and moved to its source. He was soon handed its torch and kept it burning through fair weather and foul.

As an adolescent during World War II, I found shelter in the isolated and impoverished highlands of central France. There I was introduced to a world of images through outdated math books filled with illustrations. After the war, upon acceptance into the École Normale Supérieure, I realized that mathematics cut off from the mysteries of the real world was not for me, so I took a different path.

★ ★ ★

Half a century before I was born, Georg Cantor (1845–1918) claimed that *the essence of mathematics resides in its freedom.* His peers went on to invent—or so they thought—a batch of shapes called "monsters," or "pathologies," and their study pushed mathematics into a deliberate flight from nature. Helped by computers, I actually drew those shapes and diametrically inverted their original intent. I went on to invent many more, and identified a few as tools that might help handle a host of often ancient concrete problems–"questions once reserved for poets and children."

Within the purest of mathematics, my unabashed play with abandoned "pathologies" led me to a number of far-flung discoveries. An exquisitely complex shape now known as the Mandelbrot set has been called the most complex object in mathematics. I pioneered the examination of reams of pictures and extracted from them many abstract conjectures that proved to be extremely difficult, motivated a quantity of hard work, and brought high rewards.

Within the sciences of nature, I was a pioneer in the study of familiar shapes, like mountains, coastlines, clouds, turbulent eddies, galaxy clusters, trees, the weather, and others beyond counting.

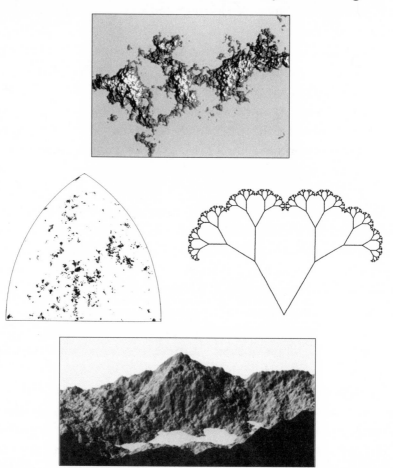

Within the study of man's works, I began with a curio: a law for word frequencies. I peaked with an extremely down-to-earth issue: the "misbehaviors" observed in the variation in speculative markets. And I added my grain of salt to the study of visual art.

So where do I *really* belong? I avoid saying everywhere—which switches all too easily to nowhere. Instead, when pressed, I call myself a fractalist. A challenge I kept encountering—one I never knew quite how to manage—was to do justice to the parts and the whole. In this memoir, I try very hard.

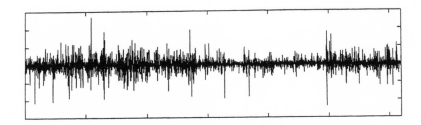

Altogether, plain old-fashioned roughness in science and art is no longer a no-man's-land. I provided a theory and showed that an astonishing number and variety of questions can now be tackled with powerful new tools. They challenge standard geometry's conventional view of nature, one that regards rough forms as formless. It appears that, responding to that ancient invitation of Plato, I have extended the scope of rational science to yet another basic sensation of man, one that had for so long remained untamed.

In a life far more interrupted than I would have preferred, basic stability was provided for thirty-five years by IBM Research and then for many years by Yale, and I lived long enough for my work to be appreciated in more grand ways than I ever imagined.

Writing this memoir earlier might have made my professional life a bit easier. But the delay has been fruitful. It has rubbed out some less important details, and my life's course has become clearer, even to me.

★ ★ ★

In this memoir, exact quotes are in italics. Conversations I recall vividly are in roman type between quotation marks. There are no footnotes and few references.

Part One

How I Came to Be a Scientist

All truths are easy to understand once they are discovered;
the point is to discover them.

—Galileo

1

Roots: Of Flesh and the Mind

IN PEACEFUL AND PROSPEROUS COUNTRIES, the children of landowners, bakers, or bankers have an easy option: to follow the family traditions and trade. But I was born in Poland and my family came from Lithuania—neither peaceful nor prosperous. As observed by a writer native to that part of Europe, *Woe to the poet born in an interesting piece of geography in a violent time.*

My ancestors' main inheritable property consisted of well-worn books. Indeed, over many generations, the family tradition has been to forsake greed and to worship works of the mind.

> *To be a scientist, a thinker, an inventor, was viewed as a higher call-ing, as something near divine. A scientist or creative mind was described as "immense." For the young people in our household and my friends it was an incredible, extraordinary privilege to be allowed to think and devote one's life to science. Money counted for little and one did not [seek] a path to riches or a career. Not at all! To the con-trary, one wanted to sacrifice oneself for science.*

The author of these words was my uncle Szolem (1899–1983). In very different ways, he and I both made this sacrifice. He became a well-known mainstream mathematician. His words may sound naïve, even corny. They strike me as describing an extraordinary alloy of Jewish and Russian traditions, in many ways a high point in man's readiness to face insoluble issues. His world would not have known the meaning of the word "corny" and that motivated many forms of heroism . . . and of destruction.

On gifted youngsters, this environment bestowed absolutely no feeling of entitlement and offered no encouragement by flattery. Not only did it not provide shelter from the tragic reality of life, but it imposed a heavy burden: to shine or at least to try to become scholars of some sort—but not without leaving time for a family and fun.

How did I react? In one word, I obliged. But contrary to Uncle and influenced by World War II in France, I took a path that nobody I knew of had previously attempted.

Momentous Dinner Party

Too often, an event's importance is not recognized until it is too late for proper recording. Also, many plays or retellings of history contrive an early scene that recalls the past and brings the main actors together. In my life, such a scene actually happened and was properly recorded in June 1930, in the family home where I was born on November 20, 1924.

This extraordinary event brought several of the most important people in my life together at the same table. Through those present, mathematical ideas rooted in the late nineteenth century would have a greater and more direct influence on my life and work than would a twentieth-century invention, the computer.

The stage was the dining room of our family apartment, at Ulica Muranowska 14, in the Warsaw ghetto. Across from a pocket-size park, it overlooked the shell of a large building that was abandoned during construction and was probably still not finished when World War II flattened that whole area, that whole world.

A professional photographer hired for the occasion produced an instant family heirloom, admired and commented upon for as long as I can remember. It documented much of my family history and the fact that I grew up in what may be called a house of mathematics.

Every person at that dinner deeply affected either my blood or my spirit. At different times, they have been examples to follow, inflexible spurs, or a stern panel of judges. Being a maverick weakly rooted in his present, I have also found them a lasting source of comfort. Let me introduce these actors, then return to the main ones at greater leisure.

The hostess and only woman in this photograph was Aunt Helena Loterman. Of Father's four sisters (including two dentists), she was the third and the only one willing and able to keep a home. Hence, in a community in which the women were formally educated before the men—and were expected to work outside the home if they could— Helena remained a contented full-time childless housewife. She was also the unflappable caregiver for the eighty-year-old white-bearded patriarch in this photo, her father and my grandfather, Szlomo, who died five years after this event.

Grandfather's only language was Yiddish—my only language was Polish—so we could not communicate. But his influence on our family was profound. He was born in a sizable old city in the Russian Empire that a cruel history had endowed with many alternative names. He called it Vilna. Poles write it "Wilno" and pronounce it "Vilno." Reclaiming an older name, Vilnius, it is again the capital of independent Lithuania, today a small Baltic state, but once a powerful grand duchy that extended to the Black Sea and became linked with Poland. Napoléon Bonaparte, on his ill-fated journey to conquer Moscow in 1812, called it the Jerusalem of the North. My ancestors on both sides had lived there for five centuries, practicing an intellectual variant of Judaism, somewhat Calvinist as opposed to the Baptist-like Hasidic variants that arose farther south in Ukraine. Economic opportunity brought Szlomo to booming Warsaw, where Father was born.

The few families sharing our name—under variant spellings—may or may not be related. But it is indeed a proper Ashkenazi construction, and John Hersey gave it to a hero in his book on the World War II Warsaw ghetto.

Reportedly, all of Father's male ancestors belonged to the caste of priests and were men of great learning, some even famous within Jewry. Following tradition, each made sure that his preferred daughter married his preferred disciple; that is how Grandfather's teacher became my great-grandfather.

As proclaimed by their distinctive styles of dress, a sharp break had occurred between the generations. Grandfather and other elderly relatives belonged to a ghetto where religion was paramount. Their children belonged to an altogether different world, where religion mattered far less. We never felt rich, but Grandfather's household seemed comfortable and sometimes had one or even two peasant servants. How did he manage before his many children could support him? I never could help wondering. Ostensibly, by purchasing yeast wholesale and retailing it to regular customers. Szolem, his youngest son, made the deliveries when he was a boy. But there was more. That community strove to support its learned men. Grandfather had been a respected and beloved adviser. For the more prosperous men praying with him, commercial dealings were a nice cover to make him comfortable and put him in a position to hold court—ensuring access to his valued advice.

Leaning on the back of Grandfather's chair is my cousin Leon (circa 1900–70), then an editor at the most important Polish-language Jewish daily. He kept us all in touch with what was really happening. He was to escape the war in Poland by being deported to eastern Siberia, then moved back west again. We met several times after the war. His wife, Maria Bar, was a leading pianist; I saw posters of her concerts but never heard her play. Leon's brother, Zygmunt, was a schoolteacher and well-known poet.

The thoughtful forty-seven-year-old man facing the camera second from the left is Father (1883–1952), second of Grandfather's four sons, a deeply principled and fiercely independent man and a major figure in my life. Two older siblings married young and joined their spouses' families far from Warsaw—but not Father. Selfless as a son, brother, husband, and father, he held on until very late in life before he finally complained in front of me that outside of his family he had done nothing he enjoyed, while his youngest brother, Szolem, had never done anything but.

Szolem is the thirty-one-year-old man at the far left of the photograph. Grandmother was fifty at the time of her sixteenth pregnancy; she never quite recovered and died before World War I. Intellectually and financially, her last child was largely raised by Father.

At that momentous dinner, Szolem was both a host and an honored guest, and probably the main interpreter. As a child, he had lived at that very same apartment, and he was the first in Father's family to attend an academic—rather than a religious or trade—high school and a university other than medical. He moved to Paris, where he had been rapidly and thoroughly accepted, and was visiting his birthplace in triumph. He was one of four professors on the way to represent France at a big event to be held in June 1930 in Kharkov, in eastern Ukraine, namely the First Congress of Mathematics of the Union of Soviet Socialist Republics. No one would influence my scientific life as much as Szolem.

At the place of honor to Grandfather's right sits the most senior guest, Jacques Hadamard (1865–1963). As far back as I can remember, I heard him described as a great scientist, arguably the greatest mathematician in France at that time and one widely believed to deserve much broader recognition than he now has. In different ways, he also turned out to have an enormous influence on my life in very specific and important areas. I think of him as my grandfather of the mind.

To Grandfather's left sits mathematician Paul Montel (1876–1975), who only a few years before had supervised Szolem's doctoral dissertation. I interacted occasionally with Montel, after a time when he had become rather remote. Szolem praised the work of two mathematicians directly inspired by Montel, Gaston Julia (1893–1978) and Pierre Fatou (1878–1929). Around 1980, I had the privilege of joining Montel's scientific progeny by discovering the Mandelbrot set, the work that made our family name known far and wide. It revived sober words and formulas of Montel's school in the 1910s and transformed them into both active scientific research and continuing infatuation among the young. It even created a popular fad.

The mathematician Arnaud Denjoy (1884–1974), who sits in the center, only affected one corner of my work, not an important one.

Finally, next to Aunt Helena stands her husband, Loterman, handsome, sweet, and cultured. As my private tutor, he was to help mold a very peculiar "formal" education. In Warsaw at that stage of the Depression, regular employment was a privilege that neither Loterman, nor Mother's younger brother, nor many other relatives ever seemed to enjoy. But strong family solidarity somehow saved them all from menial jobs. The Lotermans vanished in the Holocaust.

Father

One guest at that momentous dinner demands a closer look. Father was sixteen when Szolem was born, attending a trade high school, where he learned to be a book-keeper. He was a lifelong self-improver and an extremely widely read, clear-minded, and scholarly person. He was fascinated with how machines worked and very handy with tools—in wartime, he taught me many tricks.

A man is known by his heroes. Next to the lens grinder and philosopher Baruch Spinoza, Father held special admiration for Charles Proteus Steinmetz, a cripple who became a prolific inventor and an advocate for competent and honest government in Schenectady, New York, where he worked. Father's creativity in mathematics was never tested, but he had a phenomenal practical gift for numbers: when adding two-foot-long columns, his sharp pencil just flew down and up and he never made a mistake.

An episode during the German occupation of France illustrates his intelligence, independence, and daring. At one point, when he was imprisoned at a collection point on the way to the death camps, a Resistance group broke in and overcame the guards. They shouted that they could only open the gates but not defend the camp, told everyone to run, and disappeared. Father found himself in a long line of prisoners walking toward the nearest city. He sensed trouble, switched to a side road, and watched in horror as a Stuka plane of the Waffen SS—alerted by the guards—strafed the prisoners. Walking home, he kept to back roads, sleeping in abandoned shacks along the way. Other war survivors describe being in a herd on the way to the death camps, noticing a way out, and taking it instantly. That is the kind of man Father was.

He was a reluctant businessman forced by fate into professions related to clothing—the "needle" or "rags" trade. This activity brought him no fulfillment and played a small role in my life. He did not train me in "business wiles." Never an "organization man," he always managed to be on his own, or at worst with one partner.

Among my earliest recollections are visits to his wholesale business in ladies' hosiery: stockings, tricots, and gloves. His shop was located at Ulica Nalewki 18, a major shopping street in the Jewish quarter, on the ground floor in the back of what seemed to me a big courtyard. World War II turned Warsaw to rubble, and Nalewki was rebuilt as a short paved street along a park, where no one lives—a mere shadow of its former teeming self. Recently, a well-wisher mailed me a copy of an old trade directory showing that Nalewki 18 housed a high proportion of similar businesses—proof that Father had chosen the best possible location in Warsaw. His business was one of the few to be "registered" (whatever that meant), to have a telephone, and to be listed in boldface letters. He had done well.

The door from the street to the courtyard was always "guarded" by beggars. Father's regular suppliers and buyers often had to stay overnight with us because, although Warsaw had palaces and flop-houses, it had no affordable business hotels. Father's business moved on confidence and credit, and both collapsed a year after that momentous dinner. I recall vividly a visitor who wondered what happened. Mother brought in and opened a big suitcase filled with copies of invoices: "None have been paid because everybody is bankrupt; that is what happened."

Unbent, Father went to Paris in 1931 to seek a better life and brought his family to join him in 1936. Having escaped Poland, he also attempted to escape the needle trade for activities closer to his personality and ambitions, including trying to be a freelance inventor. One of his gadgets, which he called *Géographie amusante "Terra,"* even received a patent. However, Paris too was affected by the Depression—though less horribly than Poland and the United States—and he could not make a living that way. Soon he was forced to be realistic and became the junior partner at a tiny manufacturer of cheap children's clothing.

After the war, he found a job as an accountant for the U.S. Army. Mother argued that, having passed sixty, he should play it safe, give up independence, and take a quiet salaried job, at least "until our situation settles down." As she aged, this became a favorite phrase of hers—though I never tired of reminding her that their "situation" had remained unsettled since 1914.

Instead, Father started yet another new business, an amazing feat that circumstances made even more difficult. It was accomplished, almost single-handedly and on a very limited budget, in a tenement far from the garment district. He ordered his cloth from old-fashioned cloth makers in distant mill towns, then cut it himself, sometimes with my help. His margin would disappear if too much cloth fell on the floor as worthless rags, and at that time stylishness was of low priority. The actual tailoring was outsourced to homebound housewives in some outlying suburb. The son of one of the seamstresses, a truck driver, moonlighted as Father's dispatcher.

Then Father became his own salesman. He traveled alone to little towns and sold his goods directly to mom-and-pop merchants who wandered from one country fair to another—people of an altogether different culture. He had once visited those merchants, and he reminded them after the war that he was available, punctual, and inexpensive. Practically all his customers and suppliers returned.

Father's boldness worked. He did well enough to move up close to the proper garment district, buying an apartment/workshop in the declining hat district (a neighborhood currently favored by Kurds from Turkey!).

As Father was fighting his final illness, increasing general prosperity destroyed his niche profession. When Mother was not watching, I sold cheap the remaining rolls of low-quality wool and packages of unfashionable clothing, then told a charitable lie by boasting of having made a killing. This was a lesson in real economics, in how elusive and how quickly changing the notion of monetary value can be.

Mother

When that 1930 picture was taken, my immediate family was avoiding the oppressive heat of Warsaw by summering in Świder, a bathing spot on the Vistula River. Most days Father was in town for business, but Mother stayed with us, which is why she did not take her rightful place in that momentous dinner party.

Here she is at three stages of her life: a young woman in a formal studio photograph in 1935, a lioness in a 1942 identity-card photo during the war, and a relaxed grandmother in 1962.

Mother was born in a sizable town she spoke of as "Shavlee," which Poles write as "Szawli." Today it is Šiauliai in Lithuania, northwest of the capital, Vilnius. As a child, she lived in St. Petersburg, the capital of the Russian Empire, which included the Grand Duchy of Lithuania. Her family went on to Warsaw, in part because her mother was sickened by the wet, freezing winters farther north.

Born not long after Napoléon's disastrous campaign to conquer Russia, Mother's paternal grandfather had a streak of wildness, left home as a teenager, and walked to St. Petersburg. Eventually, he came back home to start a family. He held remarkably advanced ideas. As confirmed by several cousins of Mother who scattered around the world, he pioneered by insisting that all his granddaughters become doctors. At age ninety-four, he fell off the horse he was riding and died. On that horse he looked like any other denizen of the ghetto, but imperial Russia ruled its Jews in many different ways. He had met and impressed an extremely rich man named Sergei Yulyevich Witte and helped him run his estates. They went on writing to each other, even in 1905, when his former employer had become a count of imperial Russia and the czar's prime minister. That good man did not last, and his successor helped move that empire toward the abyss.

Mother was the kind of person who witnessed her world collapse around her six times, regained her composure in no time, and soon resumed full steam. Only in old age did she mention having nightmares of what she had experienced but kept to herself in her prime. Szolem repeatedly said that "she had a bad character . . . but that is often said of someone with a strong personality."

At twenty, during the time of the failed 1905 revolution in Russia, she forsook politics for study. She had a gift for languages; her Yiddish was near German, her Polish, Russian, and German were flawless, and—more important—her French became very good.

She beat the hated quota system of the Imperial University of Warsaw Medical School. In fact, she came out first, part of a brilliant generation of pathbreakers—and proud of it. She had to study the high school curriculum all by herself. The New Testament was a compulsory subject, and the textbook's binding incorporated a big cross—hence, before being smuggled into her parents' home, it was hidden

in plain brown wrapping. When terminal illness was making her lose her dignity, that was the part of her life she clung to the longest.

She chose dentistry, the medical specialty most compatible with motherhood: no night calls and fewer nasty bugs in an epidemic-prone part of the world. Before generalized anesthesia, a dentist's reputation depended greatly on speed in pulling teeth, and I recall Mother's strong right hand and powerful biceps.

This photo shows me in Świder in 1930 with my brother, Léon, fifteen months younger than I. As children, we were seldom separated, and of course we argued endlessly. To have him as a worthy sparring partner was one of the best things in my life. Small differences grew in time, and he took some turns he could not help and I could not help reverse. Sad to say, being my brother may have been one of the worst burdens in his own life.

This street scene, photographed at roughly the same time as the momentous dinner party, shows Mother to the left, holding my hand. Until I became old, bald, and fat, I did not change that much. After all these years, I can still squeeze my mind into the skin of that little boy. Unhurried, not a poet, but attentive, careful, and taking the world head-on.

The same street scene also introduces Mother's inseparable younger sister, Raya, holding the hand of Léon. Raya lived a few safe blocks away, so we could walk there alone. Having no children of her own, she was our always free and eager "deputy mother." For example, being herself a dentist, she took care of her nephews' teeth. For their offices' waiting rooms, the sisters chose the same furniture, except that the finish was black for one and tan for the other. Raya was essential to our happy and carefree childhood in a large extended family. We adored her, but her fate was to stay behind when we left Warsaw. She perished in the Holocaust.

Mother had two brothers. Her younger brother was a charming wastrel. Her older brother had moved from Lithuania to Sweden, but then came back. A fateful return! Had he stayed, he might have brought us to his new country, and our lives would have been altogether different. He reached France in 1939, but did not live long. His wife and daughter followed, then moved to the United States.

Parents, Continued

Among those with a tragic view of life who trust hard work and don't accept that anything is impossible, my parents—taken singly or together—were of championship class. Father was bold and Mother was cautious. They never shouted at each other but argued constantly about strategy, and they taught me very early that before taking big risks, one must carefully figure the odds.

Their families were of comparable social standing, Father's being higher intellectually. They met when they were children. Father was a classmate of Mother's older brother. Until both were established in a profession, they remained engaged. When Father was traveling, his daily postcard to Mother was addressed to "Szanowna [esteemed, honored]" Miss Lurie. Through endless moves, Mother managed to preserve those cards as a private treasure. But at one point, Léon and I found the package and tore the cards to keep the rare stamps. Mother sobbed and I am still ashamed.

They finally married just after World War I. Photographs show their first son as quite handsome, and everyone recalled him as extraordinarily gifted. He died in an epidemic of meningitis. Mother was so distraught that Aunt Helena had to hold the dying child. Until close to her own death, whenever she thought of him, Mother cried. Two sons born after that loss diminished her grief, but increased her expectations. All of that contributed to the ferocity of my parents' love for their two children.

My high level of self-confidence had its roots at home. It was nurtured at an early age. Both parents worshiped individual achievement, but because of the Depression and the war, they never achieved what they wanted and deserved. So their ambition and high expectations were transferred to me. Actual achievement came later. It took a long time for me to evolve into the image they implanted in me—and perhaps to even fulfill their expectations.

When World War I erupted, my parents and their first son were living in Warsaw. In their families, Germany was an admired beacon of civilization, Russia (except its musicians and writers) was despised, and France and Britain were thought too far away. But Father's business was ruined, so my parents moved to Kharkov. They lived there during the gory civil war that followed the Communist takeover in Russia, during which control alternated between equally ruthless Reds and Whites. Ruined once again, they managed in 1919 an acrobatic escape—south to Sevastopol in Crimea, then west by sea to Constança in Romania, and north back home to Warsaw. They reestablished themselves, only to be ruined a third time by the Depression, a fourth by World War II, and a fifth and final time by a nonpolitical event, Father's cancer.

Nuances of my parents' Lithuanian roots mattered in many ways, important or simply pesky. For example, in 1919, the newly reunited Poland tried to rebuild the old dynastic union. It was rebuffed, yet annexed southeastern Lithuania around Vilnius, the historical capital—but not Mother's birthplace. An armistice was in force but peace was never signed. Letters from Mother's older brother in Lithuania had to go through a business partner in Danzig (today's Gdańsk), which was then a free city. Far more serious was the fact that the armistice made Mother an "enemy alien" in Poland—an illegal immigrant. Appropriate bribes saved her from being expelled back to a country she did not remember and away from her family and friends. But our later move to Paris brought an incidental minor delight. To have been born in Šiauliai rather than Warsaw became safe. Between the wars, Jews of Lithuanian descent residing in Poland were citizens in theory but in fact were viewed as foreigners in two undesirable ways. Moving to France replaced both ways of being a foreigner with a third and far less undesirable one, and moving on to America brought a fourth, again a very different one.

In my case, things were much better in France and America than in Poland, but the onus of remaining a foreigner persisted, expanding from countries to fields of science. This did not prevent me from functioning well enough. But even for an accomplished foreigner, repetition does not make uprooting any easier. It carries a heavy price.

Uncle Szolem

Recall this book's dedication. Along with my parents and my wife, Aliette, Uncle Szolem is one of the four people who had the deepest and broadest influence on my life. The love of his mind was mathematics.

As a teenager, he started attending university courses and became familiar with "modern" concepts that were about to be organized into the Polish mathematics movement. During the civil war that followed the Russian Revolution, he spent a short time in Kharkov, which had an immense effect on the rest of his life—and mine. Attending the university lectures of the mathematician Sergei Bernstein (1880–1968), a fresh Ph.D. from Paris, he fell in love for life with the works of Poincaré and his intellectual heirs who dominated the scene in Paris. Returning to Warsaw, Szolem witnessed Polish mathematics arise as a militantly abstract field, was repelled, and went to France—a refugee driven by an ideology that was almost purely intellectual.

The fact that my parents, as economic and political refugees, joined Szolem in France saved our lives.

Years later, on a kind of "wall of honor" in his Paris study, Szolem hung a photograph that his mentor, Jacques Hadamard, had dedicated to him as his "spiritual son." Hadamard had spent most of his working life as professor at the Collège de France, an ancient and famed postgraduate institution. In 1937, Szolem succeeded him in that chair. In 1973, Szolem was elected to the Académie des Sciences to a chair held by the great scientist Henri Poincaré, then for a long time by Hadamard, followed briefly by Paul Lévy (to be introduced in due time).

A record of Szolem's brilliance and the reasons for his being quickly and widely accepted is found in a letter dated August 28, 1924, from Paul Jouhandeau to Max Jacob—one very well-known French literary figure to another. The words that follow fit the look of a more or less contemporary photograph.

[I met a] mathematician of genius who revealed mathematics to me. He says that they are the same thing as poetry, that one invents mathematical beauty, and that true mathematicians never do arithmetic. Those who invent formulas that are important and revolutionary renew science and bear no resemblance to calculators. This man is a Pole and a genius; he walks around with letters of recommendation from the world's greatest scientists and shows them with childish pride. He is in love, blond, brutal, and has the most beautiful eyes. He draws with equal genius and has never learned to draw. Sometimes he becomes madly gay and describes people with admirable satirical feroc-

ity. He is a Pole but with something from the Tirol (la la la-iou)
and makes me think of officers who fight duels in the Caucasus . . .
I think he is an immensely good person and capable of unheard vio-
lence if forced, but it may be that only the language is "violent."

The last words suggest that Jouhandeau was a fine judge of people. He might have added that Szolem carefully insulated mathematics from pictures. I made them work together for wide benefit. This
difference was fated to become a bone of contention between us
throughout my life.

Szolem's timing was perfect as mine was a generation later during
IBM's heyday as a scientific powerhouse. After the carnage of World
War I, Hadamard and Paul Montel recognized that fresh blood was
desperately needed, and were delighted to find themselves a successor who closely shared their interests. Therefore, Szolem encountered
open arms rather than competition or discrimination. Later, many
foreigners flocked to Paris. Competition revived and discrimination
returned. Like Poincaré and Hadamard, and Isaac Newton long
before them, Szolem viewed mathematics as almost real, but with a
crucial difference. They were fascinated by profound issues of physics
and the actual world, but Szolem was not.

He befriended a brilliant and driven younger man, André Weil
(1906–98), soon to become the founder and forceful leader of a new
generation of French mathematicians who emerged just after World
War I. Szolem was invited to join Weil's circle, and they cofounded a
mathematical "secret" cult that called itself Nicolas Bourbaki. The
original title of their book—*The Fundamental Structures of Mathematical Analysis* signaled their expectation that analysis would remain
among the topics "approved" by Bourbaki.

But this did not happen. After spending World War II in the
United States, Szolem returned to Paris. Bourbaki was coming to
power, and had narrowed and hardened, putting Uncle in an incongruous and uncomfortable situation. He had survived by fleeing the
Polish ivory tower, only to fall into the French one. When pressed, he
tried to distinguish between abstraction for its own sake and abstrac-

tion for the sake of the future—a distinction that was lost on me. He remained personally grateful for his Bourbaki friends' early welcome and help, and he deferred to their taste when his vote was needed. But the conflict between true love and friendships persisted to the end of his life. He belonged in the old ivory tower—a fact that will matter greatly to me.

One last comment: although deeply devoted to mathematics, Szolem found enough leisure time to join several literary and political avant-garde groups of Roaring Twenties Paris. He befriended other brilliant immigrants who kept the inner fires started in Eastern Europe burning, but he adopted French ways very quickly and soon sharply diverged from the immigrant group. Those friends published short-lived periodicals with timeless titles such as *Philosophies* and *L'Esprit,* but also *La Revue Marxiste.* He and I never discussed Marxism, and he recalled horror stories about the USSR. Several of his friends, however, were serious about radical politics and perished during the war: Georges Politzer became a pro-Soviet Communist leader; Paul Nizan later moved in the orbit of Jean-Paul Sartre (1905–80). Another friend was philosopher Jean Wahl (1888–1974), a pillar of the Sorbonne. Szolem's literary friends were the precursors of a group centered on Sartre that became far better known after 1945, the existentialists. Periods of intellectual ferment mix aristocrats and penniless immigrants.

Intellectual Dynasties

When Hadamard took Szolem as his young protégé, his daughter Jacqueline was of compatible age and unmarried. Therefore, Szolem's marriage to Gladys Grunwald countered a well-established custom.

Szolem's Ph.D. committee chairman, Émile Picard (1856–1941), had married the daughter of his mentor, the brilliant Charles Hermite (1822–1901), who in turn married into the family of his mentor, Joseph Bertrand (1822–1900). Helped by family connections, those individuals of varying levels of achievement lorded for generations over the politics of French mathematics. Orphaned when still very young, Gladys became accustomed to being asked about her father's health and responding that Monsieur Hadamard was fine or perhaps had the flu.

The social custom in question persisted. Peeking ahead, it led some to expect that I would marry Hadamard's granddaughter or perhaps Paul Lévy's grandniece. Also, the alumni of the school I was to attend held regular dances to introduce their daughters to up-and-coming recruits. I went to that "market" once but chose to follow Szolem's example of "exogamy" and married Aliette. Like many social customs, it could be defied, but at a cost: not being part of a system of patronage that is pervasive in intellectual and professional groups. Hadamard remained Szolem's patron, but my "disobedience" surely contributed to my never acquiring one.

2

Child in Warsaw, 1924–36

Now that I have introduced the key actors around Grandfather's dinner table in 1930, let me turn to *my* story. A tree's roots are important, but less important than its fruit, and describing them is slippery territory. With age, even half-successful people favor family and social friends over truly formative events. I shall try to be fair to both.

Large Family and a Carefree Childhood

The only apartment I recall from Warsaw was at Ulica Ogrodowa 7 (Garden Street), a treeless, straight, and charmless side street off Ulica Solna (Salt Street). The neighborhood, close to the Jewish area, was quiet, with one exception. Warsaw was often crossed by processions of people carrying banners proclaiming support for one cause or another. For some reason, the police pushed protesters into the block where we lived, then rushed toward them with truncheons. We watched from the safety of our balconies, rarely knowing what was happening—but clearly seeing that the political situation was unstable and ominous.

A fourth-floor walk-up, U.S. style, was the farthest up a moderately high-class dentist's patients would climb. Mother was a dental surgeon, and the elegant street side was reserved for her surgery and a nice waiting room, both heated by one large through-the-wall stove lined with white and blue porcelain tiles, like in Old Dutch paintings.

The living quarters facing the back court-

21

yard were more austere. The kitchen was as far removed as possible to lessen the smell of boiling cabbage. The ceilings were high, a valuable luxury in the heat of summer, and the kitchen had a mezzanine for our old cook and maid, Boniusiowa. We wore custom-made shoes—a sign of prosperity, but only relative to the cobblers' notorious poverty. Later, when Father left to find us a place to live in Paris, Boniusiowa had to go, and the off-the-street half of the apartment was sublet. Mother and sons moved into the former waiting room; the patients who had to wait were relocated to the rearranged entrance hall.

The fairly large bathroom was very important. Warsaw's polluting horses, dust, and dirt were not up to Mother's health standards. So Léon and I were constantly washing our hands. Each time we returned from the park during the hot summer, we stripped down and took a freezing cold shower.

During the Depression—long before any medical insurance— Mother's dental practice was excruciatingly slow. Patients came only when the pain was unbearable—with one memorable exception. One morning at seven, the doorbell rang and a young man stepped in, accompanied by an overwhelming stench of manure. He apologized for coming straight from the slaughterhouse, where he had taken his charges, and explained that his beloved would not kiss him because all his teeth were rotten and his mouth smelled. He wanted this fixed, had enough money to pay, and also brought along some fresh meat. The apartment had to be thoroughly aired after his visits. But times were tough. For a while, the demands of one patient's beloved paid many bills.

Few early childhood recollections can be dated precisely. I still see myself in my mind's eye endlessly walking through Warsaw and playing in one of its beautiful parks. The Saxon Garden (Ogród Saski) was a memorial to Augustus the Strong, hereditary king of Saxony and elected king of Poland.

I remember my initiation to the mystery of the value of money. I observed, or was told—by reports of the truth or subtle pedagogy— that a one-kilogram chunk of farmer cheese cost one zloty, a silver coin, then and now the name of the basic unit of Polish currency. But one kilogram of butter cost far more. Also, the price of fruit varied as the quality ranged from perfect to rotten. That is, long before I heard of the gold standard, I had relied on the farmer cheese standard. Cheese and butter were basically well-defined concepts. When we moved to Paris, I vividly recall Mother being flabbergasted by the variety of foods available even in the slum where we lived. The "thousand" different kinds of French cheese were a cliché, but butter! All the way from economy to premium and then superpremium: Isigny butter.

One summer, Szolem came to Warsaw with his bride, Aunt Gladys. They could not stay with us because Léon and I were quarantined with scarlet fever—the sole childhood affliction Mother had failed to shelter us from. I have a memory of them—or perhaps only Szolem—looking at us from a distance while Léon and I were engaged in a joyful pillow fight. The earliest date would have been the summer of 1927. In 1929, we were two healthy boys spending the summer in the park.

First Stage of a Rather Peculiar Education

Achieving fluency in reading and writing came early and painlessly and left no memory. Polish spelling is supposed to be phonetic and easy, but of course is not; yet I don't recall any problem. My next recollection bears a built-in date. I can see myself starting a letter by writing "January 1929," then realizing that the new year had come and gone and changing "1929" to "1930." I was five years and a few weeks old, far too young for school. Even today, I still sometimes begin a date with "19" instead of "20."

That letter was not written at home, and the error was pointed out by sweet and cultured Uncle Loterman. Starting early, and until I went to real school in the third grade, Uncle tutored me in the apartment where I was born and where that momentous dinner party was held in 1930. Officially, Mother feared epidemics. I am sure that Uncle was paid, and it may be that Mother sent me to public school when our money ran low.

A loving tutor is wonderful, but Uncle's lack of experience, organization, and teaching technique marked me for life. He was a chronically unemployed intellectual who—unlike other men we knew—did not escape idleness by earning several useless doctorates. He despised rote learning, including even the alphabet and multiplication tables; both cause me mild trouble to this day. However, small countries breed broad curiosity. He made me a skilled speed-reader. We discussed my readings, and—alas—the current events were seldom boring. He told stories from antiquity and trained my mind in an independent and creative way. We played chess constantly. Maps filled his household, like Father's, and I read and memorized them. Certainly, these experiences did no harm. For as long as I can recall, I have viewed dates and numbers as aligned over an endless mental line. Who knows, it may be that the chess and the maps helped me develop the geometric intuition that was to be my most important intellectual tool when I became a scientist.

This tutoring was to be the first stage of a peculiar education that was pushed here and there by the catastrophes of the century, alternating chaotically between short periods of relative "normalcy" and long ones of disorder. I became proficient at some things, but in many formal ways I remained extremely underschooled, both in class and in real life. Fortunately, the gaps in my formal education proved less deadly than feared.

The Second Polish Republic, an Early History Lesson

My family and friends took it for granted that events both past and present could have serious and immediate effects. Therefore, historic events were continually discussed, and I listened all the time, organizing everything in my mind. Today's small Lithuania was at one point a major Catholic imperial power that stretched from the Baltic to the Black Sea, through western Ukraine. In the Middle Ages, a dynastic union joined and partly merged the large Grand Duchy of Lithuania and the smaller Kingdom of Poland—two Catholic bulwarks against Eastern Orthodoxy. Somewhat later, the union called itself a commonwealth under a king elected by unanimous vote of the gentry. A single nobleman could derail the vote, and candidates to the throne needed deep pockets. So, many Polish kings were Hungarians, Saxons, or Swedes.

A king of Poland and Sweden replaced the old capitals of Cracow and Vilnius with Warsaw, then a small harbor at the highest point on the Vistula River that was accessible to seagoing vessels. By European standards, Warsaw—like Berlin and Madrid—has a short history and, thus, few historical landmarks.

The Polish Commonwealth lasted longer than expected. But in 1772, again in 1793, and thoroughly in 1795, it imploded—without military conquest—and was partitioned among the three empires along its borders. To the east, the few previously established Russian Jews—including a small number of protected merchants of the first guild—were joined by masses of former Polish Jews who somehow threatened the czars. Therefore, the prepartitioned eastern border of the Polish Commonwealth survived until 1917 as the eastern limit of the Pale of Settlement, where Jews could legally reside. Russian anti-Semitism arose only after Poland imploded.

After the Allies defeated Germany and Austria in 1918, and then Trotsky's Bolshevik army in 1920, Poland regained independence, with expanded and nonhistorical frontiers. For these and other truly unavoidable reasons, Polish history from 1919 to 1939 was rough. Its western frontier reached the sea via a corridor splitting Germany. In its eastern lands the upper class was largely Polish, while most inhabitants were hostile Lithuanians, Byelorussians, or Ukrainians. Completely alienated Gypsies were very visible in their typically colorful clothing. Altogether, the new Polish citizens hardly knew one another. The ethnic Poles, especially the relatively spoiled former Hapsburg subjects in the south, were disappointed that unity brought no harmony.

Last but not least, as far as we were concerned, Poland had been destroyed by Asian invasions and found itself without a middle class, so Germans—both Christians and Jews—were invited. In the 1920s, about 10 percent of the residents of Poland were middle- and lower-class Jews. Some continued to wear medieval caftans in diverse styles proclaiming their sect.

After Polish reunification, around 1920, Ignacy Paderewski, the famous Chopin pianist and short-term president of Poland, officially declared Jews the sole cause of every economic and social ill. This was the Second Polish Republic we came to know.

Unemployment was widespread, especially after the Depression hit, and lasted until the war. Emigration was high. Ethnic Polish peasants rapidly became French miners. Father's youngest sister, Regina, married a man who shepherded trainfuls of Jews from shtetls straight to Bremerhaven and then to steerage in waiting steamers headed for the United States. His plan to take the last ship himself was foiled when the United States set a tiny quota from Poland.

Ferocious nation-building was turned on its head. Earlier attempts over a hundred years had tried to transform a hodgepodge of underdogs into proper Russians, Austrians, or Germans. These were replaced by efforts to see them either leave or become Poles. One day, my elementary school teacher received orders to read aloud an official statement, which became part of our textbook. "Poland is a happy multinational country where all the ethnic problems of the past have been solved." She then looked at the class and—in effect—winked. We all knew what she meant.

Neither Poland nor its diverse citizens managed very well. Hardly any country managed—or manages—much better. Ethnic cleansing was tried and did not work. Since diversity cannot be avoided, one may as well like it (as I came to) or at least learn to live with it.

Polish Elementary Schools

In 1919, a burning desire to reestablish lost national unity led Poland to build a strong system of compulsory elementary education. In Warsaw, the schools were segregated by religion—programs were identical except for special lectures by a priest or rabbi. The Jewish schools did not teach Hebrew, which I "studied" at home unsystematically, and therefore never learned.

While exceptions to segregation were rare, the new Polish educational system that was set up by intellectuals and influenced by fashion and dogma ruled that a child should not be humiliated. Hence, after illness held Léon back for a year, he was transferred to the nearest public school, which happened to be for Catholics.

Poor and rushed, the ambitious authorities took all kinds of shortcuts in converting available space. My Public School 24 for Boys linked together two apartments on a rather high floor (I seem to remember the sixth) of a walk-up that filled a whole block above a smelly wholesale fish market. Class sizes had to fit room sizes. One room with no desks was used for breaks that classes were forced to take in turns.

The new Polish education viewed the teacher as a surrogate parent, so the same person taught everything except religion and gym and was "promoted" every year with her class. After that class graduated, the teacher was then "demoted" to the first grade. The four years I spent in the class of Mrs. Goldszlakowa were one of the "normal" periods in my schooling, which alternated with highly "abnormal" ones. They were a breeze, a pleasant experience that left few memorable impressions.

Long Summer in Belarus

At age ten, I spent an unforgettably exotic summer in an area that was then eastern Poland and is now near the center of independent Belarus. Soon after my arrival, I was warned sternly, "Watch out. If you walk toward the sunrise, you will soon reach a big wooden wall interrupted by mirador towers, with clear shooting lines. Keep away. Sometimes there are soldiers in the miradors, and they shoot without asking." So I had a first glimpse of the feared Soviet Union from across a calm meadow that had been casually split by Western diplomats and Soviet commissars. I kept away.

★ ★ ★

The ukases (laws) of imperial Russia began with "We, Czar of all Russias, King of Poland, Grand Duke of Finland . . ." One czar ruled over three distinct Russias. The Byelorussian language was also called White Russian (no relation to the Russians who fought under the white imperial flag and lost the civil war against the Red Russian Communists) and was fairly close to Polish, my native and at that time only tongue. It was even closer to Little Russian, now called Ukrainian, and to the standard Great Russian that Mother (a Russian university graduate) spoke with some reliable friends when the Polish language police were not listening.

Belarus had once been part of the Grand Duchy of Lithuania. It merged with Poland and then became part of the Russian Empire. The 1921 Treaty of Riga divided it and Ukraine between Poland and the Soviet Union, leaving Minsk, the capital, barely to the east of the border. Today, with some reluctance and for the first time in history, it is an independent country.

Its meandering rivers and deep marshes are an obstacle to both conquest and progress. My summer there was spent in a hamlet of a few farms on more or less flat land, with a bridge and water mill on a small river. The hamlet's Polish name was written "Połoczanka" and pro-

nounced "Powotchanka." The nearest little town was called Raków, pronounced "Rakoov." A map from the wonderful collection at Yale's library shows it at the very edge of the world, the meeting point of two then-Polish provinces and Soviet Byelorussia.

My generous hostess was a Mrs. Goldberg, whose sister, Mrs. Wigdorczyk, was Mother's best friend in Warsaw. This made it a safe escape from the city's summer heat and dirt. Mr. Wigdorczyk was going east on business and made a detour to escort me. Why did Léon not come? I don't recall.

Up to Wilno, the modern Paris-Leningrad train used the standard-gauge tracks of Western Europe and America. In Wilno, we transferred to a nineteenth-century wide-gauge Russian relic filled with big old-looking women carrying big heavy bundles. The map showed tracks continuing to the Soviet border, then to Minsk and beyond, but at that time passenger traffic terminated in Mołodeczno. There, the tiny remainder of Napoléon's Grande Armée that had escaped from Moscow in 1812 was hit by the lowest temperatures (–30°C) of that thoughtless adventure—as I noted years later on Minard's classic graph of its thinning ranks.

A horse cart waiting for us represented an exotic world, familiar from illustrations in old Russian novels. We drove on and on, deeper and deeper into the forest and—so I felt—the past. Finally, we reached a real old-style izba, or cottage, also like in those novels: very low ceiling and thatched roof, half buried underground, tiny and few windows to protect against the harsh weather. An enameled plaque like a city street sign gave the farm's number (I still remember it was 24), the tenant's name, and the date he moved in.

Most locals spoke Yiddish, Byelorussian, or both. For a speaker of Polish, basic Byelorussian was not hard to learn, so I could describe to everyone the wonders of Warsaw and also follow local gossip. The farm belonged to a Polish aristocrat, the "little count," who was rumored to own one hundred farms and live in Warsaw. Mr. Goldberg was literate, and to some extent he acted as "his" count's representative.

Having been told that we had driven forty versts from Moło-deczno, the little nerd I was could at long last find out the real length of a verst in meters. Schoolbooks said "about a kilometer" for good reason, one that the locals soon explained to me. Given the nature of the roads, there was a summer verst and a winter verst. Both measured time, appropriately so. Like in the Wild West of the United States, one might be stuck in a rut for a long time, so one began travel by choosing a rut carefully.

At different times, the dusty hamlet square filled with farm animals moving around without becoming lost. A young bullock and a cow taught me about the birds and the bees. Equally vivid in my memory is a fellow who moved from farm to farm to neuter the piglets. No anesthesia, no protection against germs during or after, just a sharp knife moving quickly and firmly and screaming animals running back to their mud hole. A dentist's son could only be fascinated and horrified.

Soon I could satisfy a burning curiosity. I approached a neighbor sitting barefoot on a stone wall and managed to ask him, "Why is it that you have no toes?" "Because I am old." "But so is my mother, and she has toes." "And my feet have frozen several times, and my toes fell off." His daughter, also ten, was my friend, so how old could he be?

The only continually exciting spot in the hamlet was the small water mill over a depth of about one yard of water, and the only native Polish speaker was Jósef the miller. Northern France and southern England had perfected that technology during the twelfth century, but in Połoczanka a technical expert from far away was needed.

To my surprise, that mill was not crushing grain but "fulling" wool, an activity that few have heard of, though its past importance throughout Europe is reflected in the common surname Fuller. When snowbound during the winter, the farmers spun their sheep's wool into a rough thread, then wove it into cloth with a loose square weave. This cloth was put into hot water with a basic black or brown dye. Ratchets attached to the wheel of the mill moved big wooden blocks up and then allowed them to hammer down on the cloth.

Near a wooden bridge a bit upstream from the mill, the road was falling apart. I remember my excitement when the bridge approaches were rebuilt.

The miller's girlfriend was a maid on a farm, and I was their confidential messenger back and forth. I once asked if she was happy to live in Poland instead of Russia. She responded, "Not at all!" "Why so?" "Because on Sundays the Catholic procession crosses Raków ahead of the Russian Orthodox procession. In Russia, the Orthodox worshipers are not humiliated in this fashion." In fact, our family in Warsaw knew that the popes were harshly persecuted by Stalin, but—though she lived only a short walk from that big wooden wall—she had no idea. When her Byelorussian parents found out she was involved with a "foreigner," they took her back in a rush to be married properly.

* * *

When not playing with urchins in the dust, I was roaming the fields and the forests in search of wild mushrooms.

The Goldbergs did not dare let me return to Warsaw alone, so I was still hanging around when the time came to harvest rye (wheat would not grow there). I offered to help, but it was Jewish New Year and the neighbors only allowed me to watch. A long line of stooped women from a number of farms moved across the field. Scythes were either not known or worthless on the rough ground. The sickles they used instead explained the Soviet emblem: the crossed sickle and hammer symbolized exploited farmers and workers. Men followed, pulling the rye into big bundles, then other women gathered grains that had been dropped. If one stood far enough away not to smell the sweat, it made for an idyllic postcard scene straight out of the biblical story of Ruth!

At long last, the Goldbergs selected a young woman about to leave Raków for a pioneer school in Warsaw, on her way to a colony in Palestine. She agreed to accompany me to Warsaw. We took the slow night train to Wilno. The platform we waited on for the morning train to Warsaw was almost like home for me, but she felt lost and homesick and I had to reassure her. A telegraph boy passed by carrying on his chest a box that served as a movable desk; she stopped him and paid for a wire to inform Mother I was coming. A reader and knowledgeable city boy, I was familiar with telegraphic style and insisted that Mother would understand if the telegram said "1600." But she wanted something foolproof and dictated: "I shall arrive in Warsaw at 4 p.m. Please come. Love. Your son." She was older and it was her money; how could I resist!

On arrival at the big, noisy station in Warsaw, the young woman was in full meltdown. But that stage of her ordeal was almost over: Mother was there, arranged for a droshky (a horse-drawn taxi) to take her to school, then took me home. Połoczanka itself seems not to have survived the war. What happened to my friends there?

Prisoners in Their Own Country Dream of Escape

In 1930s Warsaw, the Depression was terrible and the already bad ethnic and political strife was getting worse. My rational and decisive parents closely followed events in Germany and Russia and concluded that our prospects of happiness in Poland were grim.

Around age ten—like in Paris around age twenty—I lived through a period of loud ideological activity, rife with demagogues proposing all kinds of radical solutions, magic bullets that could not miss. A child cannot make life decisions, but I knew how to listen and watch. I am sure that my choices later on were profoundly influenced by my family's attitudes and the steps they either took or did not take.

The Jews' situation in Poland was clearly desperate, but what could be done? Join one of the Communist parties whose members or sympathizers often marched down the streets? Retire into a world of prayer and hope for the best? Join one of the Zionist parties, ranging from peaceful to ostentatiously Fascist? Seek freedom elsewhere by emigrating?

My parents had every reason not to be soft on communism. They had been caught in Kharkov, in eastern Ukraine, during the bloody civil war, when the Whites and the Reds alternated in taking deadly control. Szolem lectured on several occasions in Russia—one of them leading to the momentous dinner party in 1930—and reported what he saw. We knew about the purges, though the worst ones occurred after we fled to Paris. I also remember being invited to a Zionist outing meant to proselytize to the very young. When Mother heard me report their views, she said they were plain Fascists and forbade me to see them again.

Something had to be done, soon, but every option carried a high risk and cost. My family regarded every radical solution with outspoken suspicion, and listening to all those endless scenarios and arguments marked me for life.

A High-School Freshman Saves My Life

My brilliant cousin Mirka, a year older than I, faced her own set of difficulties. The revived Poland deeply cared for elementary schools, but high schools were less essential. She placed first on the fiercely difficult entrance exam to the only suitable girls' high school in Warsaw, but was bounced from the Jewish quota by others with better-connected parents. Informed, Szolem spoke to colleagues in Paris. Letters to influential contacts in Warsaw went up, up, and up—and Mirka was admitted.

What mighty person "fixed" Mirka's admission? He was Poland's most political and powerful mathematician, Wacław Sierpiński, whose role in my life, always indirect and never planned, cannot be overestimated. Around 1920, he induced Szolem to move to France, and he influenced my work in 1970.

My turn to take that entrance exam was coming in 1936, and the number of boys' high schools was tiny. In addition, Jósef Piłsudski (1867–1935), Poland's preeminent political figure, died, and politics took a sharp turn for the worse for the Jews. A pitiful, farcical, and scary Colonel Beck became foreign minister and boasted publicly that he could outwit Hitler.

Should we uproot with no thought of return? The timing was perfect because of my age, but dreadful because Father's position in Paris was precarious and Mother would forsake her profession and income. But Mirka's experience was the last straw: Poland was not the country my parents wanted for their sons. The decision was made.

We Pull Up Roots and Move to France

Our last weeks in Warsaw dragged on. I don't recall why; perhaps the visa was not yet signed, or our tenement in Paris was not yet available. But our lease had lapsed, and our former landlord was in a relentless hurry to renovate the place for his son. An acolyte paid peanuts to Mother, dismantled what I viewed as an elegant partition, and took it away on a handcart. We doubled up in Raya's apartment/surgery on Ulica Nowolipki, and left with what seemed an immense amount of luggage—including heavy feather comforters, essential in the cold Warsaw winters but not in Paris.

This episode was not only my last major experience of Poland but the first in which raw anti-Semitism hit me directly. This recollection differs from those of most survivors and shows only that I was a sheltered child. Polish anti-Semitism, official or popular, indirectly set every parameter in my life outside the family. Before TV and with little radio, the outside world was discussed interminably, but most of the time was distant, almost abstract. Trying hard, I don't recall Léon complaining about ill treatment in his Catholic school. I remember only one instance of being jostled and insulted, when on a movie outing our class sat next to another from a parochial school.

Before everything they had dreaded became horribly concrete in Poland, my parents' bold scheme had worked. We were in the South of France, looking and sounding native, and with many loyal friends

34

among the locals. The foreign wars felt far away, except for the anguish faced by families of the war prisoners. The most significant struggle was the civil war in France between local political factions.

How did Father manage to get a visa for his wife and sons? Frankly, I don't recall, but Léon once mentioned that we benefited from a short-lived program to reunite families broken by economic hardship.

Of the people we knew, we alone moved to France and survived. Most procrastinated—until times turned awful. Only two Warsaw friends survived: Mrs. Braude who lived just above us lost her husband but came to Paris after the war with her daughter, who was my age. She called Mother and they became friends again. Others had been detained by their precious china, or inability to sell their Bösendorfer concert grand piano, or unwillingness to abandon the park view from their windows. Mother was horrified by their stories but listened stone-faced.

3

Adolescent in Paris, 1936–39

AT THE INTERNATIONAL RAIL TERMINAL IN WARSAW, family, colleagues, neighbors, friends, former patients, and mere acquaintances jostled for time to wish Mother all the best. Each left a small gift, often a box of Polish chocolate. All wished they could come along. An endless and emotional good-bye.

Overnight, I had the first of many experiences of charter trains, the precursor of charter planes, full of refugees. Cheap, but old and slow, it followed an odd schedule, and often stopped so a full-fare express train could speed by. Across Nazi Germany, it was padlocked so nobody could slip in or out.

At the Gare du Nord in Paris, Father was waiting with Aunt Fanny, his sister, who lived nearby. An emotional but sober welcome. To follow Father and provide a future for her children, Mother had given up the prestige and income of an established physician, a nice apartment, and altogether a world where she was well rooted, known, respected, and independent. At age fifty she had chosen to be a lonely housewife living in a foreign slum. This contrast still makes my heart ache.

* * *

Mother's decision was taken "cold," by rational choice, and was carried out two full years before Hitler's army marched into Vienna, then Prague, on its way to Warsaw—and Paris.

By pulling up their deep roots in a community that only a few years later vanished in smoke, my lucid and decisive parents saved us all and earned the utmost gratitude. But pulling up roots is never a natural process, even under the best of circumstances. The last com-

munity where I did not have to question "belonging" was my childhood's Warsaw.

The France that was to stamp me indelibly was about to be hit by a
whirlwind, a collapse, and a foreign occupation that dwarfed the ends
of the two Napoleonic follies (the Cossacks camping in Paris on the
Champ de Mars in 1815 and the 1871 Prussian occupation). In 1936,
France was also about to be engulfed in a bitter civil war, far milder
than but not completely unlike the Wars of Religion and the French
Revolution.

Father Shows His Family the Very Best of Paris

On our first free evening in Paris, we took the long walk all the way
from the city's downscale east side to its upscale west, ending at the
Arc de Triomphe. The small number of horses and the large number
of cars made me realize that cars retired in Paris began a new life in
Warsaw. I learned how to pronounce "Renault" ("Renoh," more or
less). In Warsaw, it was "assimilated" by being changed to "Renlaut."

On Sundays, we were introduced to wonders: the Louvre, the old
science museum, the Latin Quarter. Surely, Warsaw had museums,
but I cannot remember visiting one.

Music was not on the main program, but some time later we went
to the Odéon theater for a performance of Ibsen's *Peer Gynt*, with incidental music by Edvard Grieg—in particular, for Solveig's aria, which
Mother loved to sing.

Father's favorite painter was Titian, and every new Titian I see
(grimy at its home in Venice or shockingly clean in London) brings to
mind that first visit to the Louvre. When Greek and Roman statuary
began to be dug out and long-reigning dynasties started collecting
great art, the pope chose first, followed by the king of France, then by
English, Russian, and German royalty and amateurs. This is presumably why the Greek statues in the Vatican Museum seem unrealisti-
cally "freshly minted," while their kin in the Louvre miss a nose or an
arm, and their kin elsewhere miss far more than that.

Unlike the Louvre, the old science museum on rue Saint-Martin has no clearly marked boundaries, but merges into a faded shopping district. Its core is the former Monastère de Saint-Martin-des-Champs, a counterpart to London's St. Martin-in-the-Fields and a reminder that in medieval Paris fields began around where the Pompidou Center is today. The first bicycle (low, wooden, and with no pedals, propelled by feet pushing the road), the first motorcar (a monster powered by steam that arguably inspired the creator of thermodynamics, physicist Sadi Carnot), the first airplane to actually, if very briefly, fly (Clément Ader's batlike contraption), the first plane to cross the English Channel (Louis Blériot's)—these and many comparable marvels of human ingenuity were hidden under thick soot in this dark Gothic-era abbatial church. The institution housing the museum was founded in 1794. Its name, the Conservatoire National des Arts et Métiers (CNAM), is old-fashioned but tells it all: here the nation preserves the originals of its greatest practical thinkers' greatest achievements. That first visit to the CNAM left an imprint, and I make it a point to return there every so often in a kind of pilgrimage to my childhood.

On our third Sunday outing in Paris, Father took us to the Latin Quarter, on the academic Left Bank, situated on the steep hill dedicated to the patron saint of Paris, the Montagne Sainte-Geneviève. We saw the works: the boulevard Saint-Michel, the Luxembourg Gardens, the doors of the Sorbonne and other universities, the Bibliothèque Sainte-Geneviève, and the Panthéon—that "big statement" building. And Father made a special point of passing behind the Panthéon by that small and discreet entrance to the lodge—the Boîte à Claques—that reads

ÉCOLE POLYTECHNIQUE

in faded gilt letters. Now the school has moved away, but the sign remains. The hope that I would cross that threshold as a student was what sustained Father. He was in seventh heaven when—nine years later—I became a polytechnicien. Like the CNAM, the school dates to 1794. Long after Father died, I was a special guest of both during their bicentennials.

This reunion with Father and those introductions to the land-marks of Paris often come to mind. Once again, each time my heart aches. Who would dare begrudge a man who, compared to his youngest brother, had achieved so little. Father was introducing his wife and sons to things he perceived as being among the most admirable and promising on earth. Geographically, they were only steps away from our tenement, but culturally they lay across a broad and deep divide he desperately wanted his sons to cross.

He must have also felt the need to reintroduce himself to his sons and his wife of twenty years. For five long years, he could afford only a few visits home, and earlier he had been consumed by efforts to help his Warsaw business survive the Depression.

Every day was a brand-new beginning in every way.

Belleville, a Slum in the Nineteenth Arrondissement

Mother was not allowed to practice dentistry in Paris, so every penny was needed for the new business Father was building. Before we joined him, he found a suitable tenement in an old slum called Belleville (Beautiful Town)—a *logement* rather than an *appartement*.

Many slums' kitschy names are commercial. But this had been an old village northeast of the city center, on the sunny southwest slope of a steep hill just beyond the fortifications that, until 1860, marked the city limits. Belleville is in the east end of Paris, at the bottom in terms of prestige—and far from the promise of its travel brochure name. In Paris, like in London, the prevalent winds blow from the west, so the nice area is the west end.

The rue de Chaumont was (and remains) a small dead-end street in the middle of a beat-up area long slated for urban renewal (which came decades later). At numbers 5–7 stood a clean and relatively nice building. Having found it, Father spoke to the concierge, who was also the owner, and in effect, he was given an exam.

Reassured that we were not homeless derelicts but a middle-class, down-at-the-heels family, the owner let us rent a flat: two narrow rooms end to end, railroad-style, from street to courtyard. One was the parents' room, filled by the dining table and bed; the other was the sons' room, filled by study tables and beds. A bit later, space was made for Mother's elder brother, who had fled Lithuania; he was ill and did not live long. Our kitchen was the size of a cupboard, and on every second landing between floors, a Turkish toilet serviced four flats. We had no hot running water and no bathroom.

Other neighborhood houses had courtyards teeming with immigrants, but we were strongly discouraged from socializing there. The owner of our building liked quiet tenants who did not linger in the courtyard. We knew hardly anyone in Paris, so our social life dropped overnight from that of a big extended family to almost nothing.

Father chose a place a short walk from the Parc des Buttes Chaumont, an amazing landmark that proper Parisians from the nice west end would not visit. Its site used to be a convenient stone quarry for Paris. Exhausted and nearing collapse, the quarry had been abandoned. The metro line crossing the park ran through an underground covered bridge supported by pillars standing on solid stone. The late-nineteenth-century architects of the park turned its gutted terrain into an asset: a lake with a high island "mountain" faced with concrete and graced on its peak by a pseudo-Greek temple with a panoramic view of the best and the worst, depending on the direction. The park was kept so meticulously clean that we could play there, as we had previously played in the Saxon Garden in Warsaw. I still recall having great fun building a big dam across a "river" raging along a gutter, and then hearing some well-dressed passersby comment, "How disgusting."

The park's broad lawns had borders of partly overlapping arches, each about a sixth of a circle, and with a curious knobby surface I found mystifying. Years later, my first visit to Japan revealed very similar arches—in bamboo! The Paris designers seem to have reproduced the texture of bamboo in more durable pig iron. So the Buttes Chaumont is called an English garden, but in fact part of the inspiration is Japanese. The same is also true of the corresponding park on the Left Bank, the Parc Montsouris, right next to which, many years later, I was to purchase my first apartment in Paris. The Buttes Chaumont area had a surprisingly rural feel with several blocks filled with models for suburban houses. I imagine that standardized developments were growing and that to exhibit their wares the builders had chosen the neighborhood accessible by metro where land was cheapest.

For a while after the war, my official home address remained 5–7, rue de Chaumont, but I did not live there. Last time I passed by to check, the area immediately surrounding the Parc des Buttes Chaumont had been built up, a solid and fully gentrified island—as expected for attractive spots in Paris. The rue de Chaumont has also improved, but our old house has not much changed. Most people in the street are less derelict than those I recall.

Back to the tiny tenement. When Mother first entered it, she sobbed uncontrollably. By the next day, she had recovered control over herself and the household. Parents and sons were forbidden to speak Polish, and it worked beyond belief. Mother brushed up on her (already fine) school French and took out books from every one of the excellent public libraries in the area. (French books being mostly paperbacks, each library had them bound in its way, and the libraries could be distinguished by the style and quality of the bindings. I soon noticed that some older books were copyrighted "in every country, including the Lowlands." The reason was that, as late as the 1900s, the center of piracy for French books was virtuous Holland.) In no time, Mother wrote nearly flawless French and spoke it with almost no accent. My parents never fit the cliché of immigrants who depend on their children to communicate in the new country.

The first time Mother went out, she commented on the small number of pregnant women. Before she went shopping, rumor had forewarned her that Parisians carried unwrapped bread in their bare hands. But she was equally shocked to see meat displayed in open stalls in the heavy traffic of the narrow avenue Secrétan. It did not look very appetizing, but it was healthy and she got used to it. Due to the constant fear of epidemics in Warsaw, the shops she had gone to were far more sanitary than those in Paris. Years later, I heard of an additional reason. Before routine refrigeration, the slaughterhouses of Paris were close to nice neighborhoods. After the best meat had been sold there, the remaining pieces moved down the social ladder, joining less desirable fresh pieces. When the meat reached the slums, its travels showed.

One day, Father lugged in an obsolete multivolume *Larousse Encyclopédie,* together with decades of bound volumes of its updates. In no time, I read them from cover to cover.

When the German advance toward Paris triggered the debacle of May 1940, my parents abandoned everything and walked hundreds of miles to join their sons in Tulle where we had been sent earlier for safety. As soon as Paris was liberated in 1944, Father rushed back. Our old flat was rented, but another tenant who had stayed behind had been deported shortly before the liberation. We could have his tenement until he came back. He didn't. Cleaning a tile in the kitchen, Mother found it was loose. Hidden under it she found a replica of a gold twenty-franc coin used during the reign of Emperor Napoléon III (1852–70). Had the previous tenant taken it along, he might have purchased his survival.

French Elementary School

Arriving in France speaking imperfect French taught by unreliable Uncle Loterman, I went straight to the local elementary school for boys, located at 119, avenue Simón Bolívar, in the Nineteenth Arrondissement. Next door was a separate girls' school, and behind it was a nursery school.

The energetic and cooperative school director had decided that Léon and I would settle in more easily by repeating fourth and fifth grades, respectively. So in 1936 I started in the class of Monsieur Poupard. Halfway through that year, we were both promoted, and I moved to the class of Monsieur Leblanc. Both teachers were excellent, helpful beyond the call of duty, never to be forgotten. Of my classmates, only one went on to high school—the same as I did. A classic slum dweller named Repkowski, he was to stay in Paris and vanish during the Holocaust.

The Parisian Dialect of French

Learning to speak French proved to be an interesting challenge. I remember having to memorize many historical dates, including those of battles that Napoléon won in Italy in the 1790s. These turned out to be real tongue twisters, since in French, "ninety-seven" is "four score and seventeen." For once Mother was forced to drill me.

Working hard I rapidly became a fluent speaker of what we thought was proper French. But in fact, I learned a significantly different language. In Belleville, they spoke "Parisian," a fully blown counterpart of London's famed Cockney. For example, *marrant* (funny) and *marron* (brown), when spoken, were hard to tell apart.

In the Paris lycée where I went next, everybody spoke High French, which began as the local dialect of the province of Touraine, like Italian began as Tuscan. Later, I transferred to a lycée in Tulle, where nearly everybody had a strong southern accent. Therefore, my spoken French never quite stabilized, and I keep an accent that varies in time and cannot easily be traced.

Certificat d'Études, Spelling, and a Good Eye

After attending French elementary school for a year, I graduated by passing the dreaded exam for the *certificat d'études élémentaires.* More than five spelling errors in a dictation forced a student to either repeat the last grade or graduate with an incomplete. I do not remember making even one.

I believe that at grade school level, spelling in both French and English is largely a matter of visual memory. Two words may be pronounced identically but spelled differently, depending on the meaning of the word, such as "there" and "their." Given my chaotic schooling and my nonstandard life choices, a good eye was going to lead me far, again and again.

Each time I recall that successful exam, my heart rejoices. Lady Luck is blind and needs assistance. In 1936, my parents assisted by moving out of Poland. In 1937, I was called to assist—and I did.

Parallel French Public School Streams

In France, many decisions are made in dusty cubicles of the Ministry of Education, where the constant political concern over minor details hides the important agendas, allowing some odd policies to remain law for longer than they should. The next stage in my life would be difficult to understand without some background on the educational system in place at that time. Until 1937, there were two distinct educational tracks. While called primary and secondary, they were in fact parallel and quite separate.

The primary track started at age six with the *écoles primaires élémentaires*, like the one I attended. The earliest the *certificat d'études*—to which it led—could be awarded was in the year of a student's thirteenth birthday. Very few students continued in school, but this track did not stop there, merely narrowed down in stages to *écoles primaires supérieures*, and then to *écoles normales d'instituteurs*, which trained elementary school teachers, who did not go to high school.

The track called secondary started at age six with preparatory classes, and grades were numbered backward, beginning with an eleventh grade. The sixth grade—two years before the minimum age for the *certificat d'etudes*—began the proper lycée that continued up to the first grade. Finally, an essential extra wrinkle—a terminal grade where students had to choose between mathematics (the sciences) or philosophy (humanities).

Those two parallel streams closely mirrored social class, and transfers from one track to the other were rare, as intended—similar to ways of impeding social mobility in many other countries. This system was overhauled in 1936, when the prime minister was Léon Blum (1872–1950), a dandy who had become the moderate Socialist leader by default. He established two-week paid vacations for everyone, and his minister of education, Jean Zay, merged those parallel tracks. Zay became controversial and hated, and during the German occupation he was murdered. His intention was admirable—but breaking an old and stable arrangement is not a simple matter. The French educational system continues to be constantly "reformed."

In 1937, following a key part of my ambitious parents' plan, I moved to an academic high school, a lycée. The nearest high school, Lycée Voltaire, had only recently been upgraded from a trade school and—according to the grapevine—had underqualified teachers. The next nearest, to the west, was Lycée Rollin, now called Lycée Decour to honor a teacher who became a Resistance hero and martyr during World War II.

The education I received in two years at Rollin was bizarre but truly outstanding. In academics I was way ahead, reading and dreaming on my own, and not heavily school-bound. So the notorious rigidities of the system did not matter, and I benefitted immensely from another feature. Because university careers were very rare, the kinds of people who today would supervise Ph.D. dissertations were teaching eleven-year-olds. They lavished on me far more than any rightful share of their time, often under the thin excuse of asking me to let the class share my broader experiences. The low point concerned gym—of which I recall only a narrow courtyard where, weather permitting, we tried to learn the long jump.

Rollin soon revealed another virtue. Not only did it serve all of northeast Paris, but it was nearest to the northern and eastern railroad stations that serviced the leftist, working-class "Red Belt" suburbs without a local lycée. Many of my classmates had long commutes and levels of commitment matching mine.

For a short metro ride, the ticket cost was not negligible. To walk was cheaper and healthier. The shortest path followed the boulevard de la Chapelle, along a neighborhood called the Goutte d'Or, somewhat east of the more widely known Pigalle and even older and lower on the scale of ancient and notorious red-light districts.

Too young to be bothered by the ladies of the night (or the day), I could not help watching the scene from the boulevard—but avoided the menacing side streets. Bored after a while, I preferred to detour with my suburban friends to their rail stations, then continue through characterless streets.

A Master Guide to French and Latin—and Paris

The study of Latin began with dreadfully dull stuff: Cicero's lawyer's orations and Julius Caesar's bone-dry report on his generalship—with no mention of the million Gauls he killed or enslaved. I did not start liking Latin until we moved on to the poets and the historian Tacitus. A belated benefit from my years of Latin is that they helped me correctly coin new words—like "fractal."

My sixth-grade teacher in charge of French and Latin was Monsieur Gilbert Rouger. Not only was he excellent, but he had edited a selection of poems by Gérard de Nerval (1808–55). M. Rouger was unforgettable for an ancillary reason: his true love was Paris. Every Sunday, he would walk through an old neighborhood, and after he had finished reacquainting himself with that neighborhood, he would start afresh. He invited the kids he taught to join him in his version of the popular book *Promenades à Paris*. M. Rouger's erudition was phenomenal—far beyond anything found on maps or in guidebooks and classic literature. His lessons served me well, notably on three occasions.

In the summer of 1945, the Allies having won in Europe, the United States decided to bring its troops home fast. A huge transportation backlog ensued, hence the need to provide the soldiers barracked near Paris with something to do. So I volunteered. As an interpreter and guide, I would be paid and not only fed but allowed to "doggie-bag" enough for Father and Mother. In 1945 Paris, Spam tasted delicious. A forceful program manager tested my spoken English and credentials, took me on enthusiastically, and gave me a variety of assignments. On the monuments and their glorious hidden courtyards, I was a walking, talking encyclopedia. My first group provided a form of hazing: foulmouthed and war-hardened WACs broadened my colloquial English, but they were bored stiff by dirty old buildings and disappointed that I neither understood their jokes nor welcomed their advances. They soon dispersed to seek fun. After that, I was assigned the legal staff of the army's headquarters—Ivy Leaguers already familiar with the guidebook stuff. They pressed me on fine points and became avid students by proxy of M. Rouger. They kept me busy even when I was trying to get loose and go home. Never a dull day.

Years later, in the 1950s, when wooing Aliette, I loved to introduce her to those classic neighborhoods. My advance descriptions of what was about to be seen were intended to impress (I was unencumbered by map or guidebook) and were numbingly accurate—except on one occasion. Between my visits, an ancient palace I had announced as lying just around the next corner had inconveniently collapsed.

More years later, in 1972, Aliette, our two boys, and I rented an apartment on the rue du Regard, in Paris. I was mystified by two buildings near that short street's other end. The lessons of M. Rouger still fresh in my mind, I agonized about style. Was it before or after 1715? When Louis XIV was ancient or when his great-grandson Louis XV was a child? But of course every Parisian I knew owned a book answering these earthshaking queries. It turned out that, during a transition period between an old king and a child, not much was built. By chance I happened to hit examples that were in a then-remote neighborhood, viewed as minor, and from a rare period that M. Rouger had omitted.

Dark Clouds

Paris had reunited our family. Compared to Poland, this was a colossal improvement. But there was no reason to relax. November 11—the anniversary of the 1918 armistice—was celebrated in grand ceremony. We all went together to the Champs-Élysées to watch the traditional military parade. In orderliness and precision, the soldiers of France did not equal those of Poland—nor the goose-stepping soldiers of Germany. I still remember Father's unease and foreboding.

My parents were trained to hope and work for the best—but also to be ready to manage the worst. It soon became clear that war was coming. One day when Szolem was visiting, he mentioned that his physicist colleague at the Collège de France, Frédéric Joliot-Curie (1900–58), had revealed to his entourage that artificial radioactivity might make powerful bombs. We were sternly warned not to mention this to anybody. I complied, and this is the first time I bring up this episode.

On another of Szolem's visits, Father made a point of telling him in my presence that to survive and help his siblings, he had abandoned study and instead he became an apprentice. Must I follow the same path? Different trade schools were discussed, but Szolem did not know about them, so this discussion petered out. War broke out soon afterward.

4

Dirt-Poor Hills of Unoccupied
Vichy France, 1939–43

WORLD WAR II REMAINS in my mind like a whirlwind. Much of the rest of my life has been dominated by what I learned—or failed to learn—during those years.

War broke out in September 1939, and from mid-1940 until 1942, northern and western France (including Paris) came under direct occupation. My parents, Léon, and I spent that time in the center and southeast—an officially unoccupied rump state in one of the poorest mountain regions of France that everybody called Vichy France.

Until 1943, we lived there in the open—conspicuous but insignificant—in an austere little town of about fifteen thousand called Tulle. We were in the most literal sense saved by devoted friends of Szolem—descendents of hardscrabble farmers and teachers from a village school, who valiantly helped Lady Luck. We have stayed in touch with two of the families that helped us survive—the Eyrolles and the Roubinets, whom you will meet.

Our constant fear was that a sufficiently determined foe might report us to an authority and we would be sent to our deaths. This happened to a close friend from Paris, Zina Morhange, a physician in a nearby county seat. Simply to eliminate the competition, another physician denounced her. (Miraculously, she came back; her daughter wrote a good book about the experience, *Chamberet: Recollections from an Ordinary Childhood.*)

We escaped this fate. Who knows why? At one point, my perfect school grades presented a clerk at the prefecture with a conflict: endemic local ill will toward Parisians and other strangers versus meritocratic ideals that ran high. In that very poor region, dreams of the good life included passing tough state exams and moving away. That clerk, a classmate's sister, played God. Xenophobia lost, meritocracy won, and she deliberately misplaced my family's files.

Luckily, we did not compete with the locals and did not seem like strangers. My parents' systematic efforts at acculturation having worked, Léon and I sounded and looked almost native.

Fearful but Intermittent Storms

"Intermittence" is a word for the old quip that army service consists of endless boredom punctuated by scary, irregular, and unpredictable interruptions. During the occupation, France saw dreadful events. All too many people experienced near-continuous horrors. Vichy France was a mixed bag, and we were lucky. I recall that period as only "intermittently" stormy.

Different arms of Vichy did not know what others were doing. One arm of the state deeply disapproved of and actively harmed unwelcome aliens like us. For example, it was illegal for my father to have any gainful occupation—and he had none. However, another arm of the same state viewed us as bona fide refugees from Paris. Being unable to go back home entitled us to public welfare: tables, bunks, and other household goods, possibly rent relief, and even a bit of money. Best of all, it included free medical service with seemingly no restriction—and the doctors in Tulle lavished house calls on our slum. Additional money must have come from relief organizations and cousins in America—who could ill afford it and deserve eternal gratitude.

When I was nearing forty, my work became devoted to the phenomenon called intermittence, present in both nature and the financial markets.

The Appalachia of France

Raised mostly in cities (Warsaw and Paris), I had been much affected by the summer I spent in that little hamlet of Połoczanka at age ten—and far more by my four years in Tulle, the southern part of the province of Limousin, the egg-shaped department of the Corrèze.

50

In addition to the constant help from friends, I survived thanks to tacit complicity of the Tullois, or Tullistes. Inhabitants of Tulle had the reputation of being unfriendly to foreigners—which included Parisians and most other French people. But after we broke down the wall of distrust, they became the most generous of hosts and helped us survive the war.

In the Limousin dialect, the word *corrèze* corresponds to the standard French *coureuse* (runner) and denotes a mountain torrent. It also denotes a little town upriver from Tulle and a not-so-close railroad stop. Its inhabitants call Tulle a city of seven hills—not one less than Rome or Paris—but it is more accurately described as built on the bottom and the sides of a very long, winding, and deep hollow with several branches. Many streets go straight uphill, and not a few include endless and infamous staircases in stone or concrete. The benefit, according to legend, is that the girls of Tulle had nicer legs than the girls of the wealthier Brive-la-Gaillarde, in the wide plain downstream along the Corrèze River. *Gaillarde* means "prosperous, strong," a contrast to Tulle's nickname, *Tulle-la-Paillarde*—"the poor one who sleeps on straw."

The sober Church of Saint-Martin sits on the more or less flat

TULLE. — Vue générale

piece of land where another torrent joins the Corrèze. Some parts date to the Romance period of the early Middle Ages, but construction took many centuries. The surrounding area is also called medieval, but most private houses are probably from the seventeenth century.

In my time, a single main street switched at each meander of the river from one bank to the other. Proper locals called it Quai (where it follows the torrent), then Rue Nationale, and then Faubourg. City hall preferred other names—the one in honor of the wartime chief of state Marshal Pétain came and went—but the locals paid no attention. Since then, that main street was made one-way, and a second was dug from old paths along the surrounding hills.

The Tulle train station, downstream from the church, was hemmed in by the torrent and hills. The Tortillard, the little train that joined Bordeaux to Clermont-Ferrand, went up the Corrèze as far as it could, then backtracked and meandered up a steep hill. The small, flat area next to the station was largely filled by an armory begun in the seventeenth century to make use of the ample waterpower. Now, of course, this armory—along with most of the local industrial jobs— is gone.

We lived near the armory. In summer, the heat in the hollow made the swimmable stretch of the river very attractive, but it was a long haul upstream from the church, even after we scrounged beat-up bicycles.

What Brought Us to Tulle?

As often happened, the indirect agent of destiny was Szolem, whose first tenured professorship had been at the University of Clermont-Ferrand. The musical chairs game of university positions might have put him in more desirable locations—Normandy, Flanders, or Alsace—with incalculably negative consequences, since those rich provinces were occupied.

In Clermont-Ferrand, Aunt Gladys and Szolem befriended a colleague, Lucie Eyrolle, called Eyrollette, who introduced them to her parents, Pierre and Louise, who worshiped hard work and the spirit of self-improvement.

Near the Corrèze train station—the second stop northeast of Tulle—the Eyrolles found Szolem an inexpensive piece of land, mostly filled with rushes, and an underemployed architect. Proud to work for a university professor, he charged nothing for supervising the builder and adapted a simple box design; unimaginative, but for slum dwellers, sheer luxury. The locals called that house *la maison du chavant*—Limousin for "the house of the scientist." A neighbor manufactured canned fruit and marmalade under a brand—Valade—that survives to this day. As a welcome gift, he brought them a big box with samples of his goods.

Léon and I were Szolem's guests there for a month in the summer of 1937, when the house was brand-new, then again in 1938 and 1939. Our first visit remains especially vivid because sweet Aunt Gladys had to teach proper French table manners to her nephews.

When the war started, Szolem, insisting on being treated like Everyman, refused a desk job appropriate to his Collège de France professorship and was instead drafted as a private in an antiaircraft unit. He, Aunt Gladys, and their son, Jacques, moved to Tulle—close to the Eyrolles and where there was a high school. Shortly after, Léon and I joined them. Just before Paris fell in June 1940, my parents joined us—with slim savings, as Father's business partner had fled with the cash box and the bank balance.

Pierre and Louise Eyrolle provided a shield for us, which we learned about later. Deep local roots gave them privileged access to the local elite, including a powerful political boss whom I never met. The unsinkable Henri Queuille (1884–1970) seemed to have been a minister in every prewar French cabinet; but in 1943, when German occupation had extended to Vichy, he could no longer help. He managed a brief comeback as prime minister after 1945, when English speakers called him Kelly—pretty close.

The Eyrolles always remained our close friends. Over fifty years later, in 1992, I was the guest of honor at a centennial related to the lycée in Tulle. Yvonne "Nini" Eyrolle Péchadre, a retired teacher, was the last of the clan. Aliette and I—together with Léon and his wife, Nicole—paid an emotional visit to the Eyrolle house. It sat on a steep hill with endless staircases and several garden terraces (now connected by a bridge to a new apartment house with an elevator). Through this grand old lady, we thanked again—and for the last time—all the Eyrolles for their active friendship and bravery. Eternal thanks.

During the fall of France in June 1940, Szolem saw little action and was demobilized in Tulle. Soon after that, he moved with his family to the Rice Institute—now University—in Houston, Texas. Earlier, after getting his Ph.D., he had spent a year at Rice, and they invited him back.

The agent of his departure was the remarkable Louis Rapkine (1904–48). He had worked at the Pasteur Institute in Paris, so he knew the French scene very well. But he was Canadian by birth and hence could travel freely. He had watched as liberals, Jews, and their spouses were expelled from Germany by Hitler and networks were established to provide new lives in the West. Almost single-handedly— though assisted financially by a Rothschild—he helped many of those who had suddenly become threatened in France. This even included Jacques Hadamard. Almost all returned by 1948, and the renewal of French science was greatly aided by the persistence and ingenuity of this suddenly privileged foreigner.

When Szolem and his family came to say good-bye before leaving for Houston, everyone wondered silently whether it was not really an adieu.

Life in Tulle

Dirt-cheap lodging was found on the top floor of a small tenement on flat land down the river near the armory. Welfare for refugees provided basic furniture. Léon and I slept in the packed kitchen/dining room heated by the Franklin stove that served to prepare food. Our parents' room was heated by an open door and cooled by three walls made of plastered straw exposed to the hill-country elements. In winter, it displayed ice stalactites. There was a Turkish toilet on the ground floor, one cold-water tap in the front room, and—needless to say—no bathroom.

We settled there into a life of the most extreme parsimony, managing with a level of ingenuity that my brilliant and resilient parents—rich in experience from previous catastrophes—could contrive when forced into inactivity. Mother regained reflexes acquired during earlier periods of great scarcity. She deprived herself to let her sons grow and became uncharacteristically gaunt.

Paper was unbelievably scarce and never thrown away if it could be reused. Cigarettes and alcohol were tightly rationed and were bartered for more essential provisions like bread and cooking oil. So I didn't smoke and learned about wine only later in life. Knowing some farmers added to our food supply. Paid entertainment was out of the question. I often walked by the sole cinema in town but never saw its inside. A radio set would hog electricity—a wasteful luxury. Travel was restricted to matters of life and death.

Even when there was no networking to do, Father moved from shop to shop, hunting for bargains and unrationed food. He fixed all kinds of stuff—like broken bicycles with no gearbox. I long kept a chair he found on the street and repaired and a knife for which he replaced a broken handle. He was extremely skilled with his hands and the tools he scrounged. Watching and helping him taught me to be handy as well. He read voraciously, always taking notes. Mindful of what fate might bring next, he used battered old books to learn to write English ("just in case"); a few of his numerous workbooks miraculously survived in a leather briefcase I found years later.

At rare intervals, the occupying German army allowed small shipments from Paris to Tulle. Our intermediary was Marie Bart, a lady who had helped raise Aunt Gladys. A heavy suitcase she managed to get through seemed valuable, but it was my collection of travel brochures. Another suitcase, filled with fine china—a wedding gift to my parents that had somehow made it through several moves—opened when Mlle Bart took it on the metro, and the well-traveled china splintered into rubble. Mother just shrugged. A few pieces survived, which I still have.

Another remaining item was a thousand-dollar bill. Distant U.S. relatives—who could hardly afford it, bless their hearts—had sent it as insurance in case of emergency or a prolonged war. We had feared that the bill was a fake, but it turned out to be real. During the post-war mess, it remained as insurance, but in 1947 it switched from saving lives to saving a penniless student. I inherited it, put it into a savings bank, and later put it to good use, thankful that fate had never forced a test.

Lycée Edmond Perrier and Mlle Tronchon

Shortly before we arrived, the locally financed Collège de Tulle had become the nationally financed Lycée Edmond Perrier, named for an alumnus and noted naturalist. I recall its buildings looking rather elegant—as confirmed on a later visit. It is located on a hill reached either by a meandering road or by those endless Tulle staircases that winter covered with treacherous ice—no salt was used.

The old staff was upgraded by attrition, but was far below the standards of the Lycée Rollin in Paris. The best teachers had been part of the staff of a major lycée expelled from Alsace, which Hitler had incorporated into the Reich.

The physics program was dull—with a negative effect on physics research in France. My education in physics came mostly from books on "how things work." My gift for mathematics was noticed. I found it easy, but it did not become essential to me until a later self-discovery. I was far more fascinated by history—learned equally from books and newspapers.

My unforgettable first French teacher, M. Rouger, was outmatched by my last—Mademoiselle Marie-Thérèse Tronchon (1907–97). She guided me through French literature to a breadth and depth well beyond any high school program. She also taught me how to write—including eighteenth-century pastiche, a skill that fell victim to the adage "Use it or lose it." Professional scientific writing can afford to be poorly written—scientific communication is by nature overwhelmingly verbal, and the audience is well defined. But I wanted to write for an audience that was mixed and not known in advance, so writing skills mattered. For a long period of time, I thought I was writing English, but I was really writing translated French. Mlle Tronchon mentioned casually one day that the English teacher knew little English, but she knew it well and would make her bottomless personal library available to me.

Formal and Informal Libraries

More important to me than school was the public library, which occupied the top floor of a cheap walk-up apartment building. Its sole librarian was kind and helpful—yet preached the German cause to anybody who would listen. The collection came from a variety of sources. One day, someone from high authority stopped by. Even though the war was raging, he would drop in every so often to inspect a private reserve of books listed on some special register—probably inherited from a church or abbey. Those books had been protected from damage by readers, but not from a leaking roof—official reports were submitted, but the books were not saved.

The Benoit Serres Catholic Bookstore near the church also served as my informal lending library. I trained myself to read a paperback without breaking its spine, leaving smudges, or otherwise revealing that it was no longer new. Ostensibly, I was buying books, being disappointed, and exchanging them. The owner knew me well (everybody did in Tulle), saw what was happening, but never said enough is enough. A good man.

Several out-of-date math books came into my hands from persons who had saved them from their student years and even their parents' student years. Invariably, they included masses of illustrations of shapes that later books omitted as a matter of principle. From these outdated books, I built in my mind a zoo of shapes that was to help immensely during the winter of 1944, when I was preparing for the very difficult mathematics exams at the Lycée du Parc in Lyon.

Baccalaureate High School Examination

The baccalaureate marks the end of college, but what is college? England and the United States settled on the age bracket of eighteen to twenty-two years; France settled on twelve to eighteen. In my time, there was a written and oral exam, plus a qualifying exam in French at the end of the junior year. The French *bac* was nominally the entrance examination to the open-admission university system. Hence, it was somewhat formal and was chaired by a professor from the region's university. In my case, this meant the University of Clermont-Ferrand, and the professor who came to chair the ordeal had been Szolem's colleague. Only later did I realize that he was a noted partisan of the Vichy regime—he could have easily denounced me but did nothing of the kind.

The committee also weighed the evaluations of the junior- and senior-year teachers. The philosophy teacher disliked my constant objections, but the key teachers were very positive. The principal's comments for the junior year were, "Bound to pass brilliantly, an exceptionally gifted and hardworking student," and for the senior year, "An exceptional candidate. Bound to pass brilliantly."

To build up suspense before announcing the results, the chairman fidgeted with his papers, then read the list backward, so my name was called last. I had a summa—reportedly the first in the school's history. Léon had come to witness the occasion and to provide moral support. We rushed home with the news. My parents saw us out the window, and from the third-floor landing Father called, "How did it go?" "*Mention très bien.*" Like an echo, he responded, "*Très bien.*"

I had performed as expected. We had no party. Nothing was said. I recall vividly a pang of regret; we all understood perfectly what it meant, but my heart ached. I never saw my parents celebrate. They may have never done so, or forgotten how—certainly they did not teach me anything along those lines.

I have no doubt that Father went around quietly making sure that every person who could possibly matter knew that I was very special.

He had fallen into the habit of repeating to everyone a statement of the mathematician Henri Poincaré to the effect that in most fields a person can be trained to become an expert, but mathematicians must be born. Times were tough, and this was not a matter of bragging, but of survival. The chronicles of the war in Eastern Europe included a growing number of stories in which a would-be "butcher" is over-supplied with potential victims and a person perceived to be special is somehow spared. Father must have felt it was very bad to be overly conspicuous, but very good to be seen as rare and special. This atti-tude, which he probably brought from Warsaw, created in me an ele-vated level of commitment and ambition.

One outcome we had desperately hoped would come from this summa was immediate and stark: an increased chance of survival. It was a new ace in hand, and all that mattered was how it was going to help in forsaken little Tulle. Its nearness to the University of Clermont-Ferrand had been, for Szolem, Tulle's greatest asset, and everybody felt that I could beat the quota system. But it was too far, too dangerous, and too expensive. A classmate who went to Cler-mont kept me informed. The final examination included two very easy problems, which I saw instantly to be a single problem stated in two different ways. Apparently, few students noticed.

Lifetime Friendship with Pierre Roubinet

Pierre Roubinet and I became classmates in Tulle late in 1939. I remember distinctly how we met. The buildings of the Lycée Edmond Perrier had been converted to a field hospital, and our grade was exiled to the site of a recently closed parochial school. On the first day of school, Pierre approached me, an obvious newcomer, and we began chatting. I soon found out that he had come from a Catholic school that could not afford the upper grades and did not fear sending its ambitious graduates to the secular state-run lycée. In one of our first exchanges, we looked up the history of the French Revolution in the textbooks that our two schools used in the previous grade. The accounts seemed to involve two entirely different countries.

Pierre's tribe and mine knew each other only by reputation—a pretty dismal one to begin with. But soon after we started chatting, reputations ceased to matter. We became close friends, and still go out of our way to visit each other whenever possible and keep in touch by telephone. Below is a picture of Pierre, on the left, me, and Léon in Tulle.

When we grew up and had families, our friendship extended to Pierre's wife, Claude, and my wife, Aliette. It was also inherited by his elder son, Martin, and my elder son, Laurent. Sadly, on a bike tour, they stopped for a swim and Laurent watched Martin be killed by a powerboat. Pierre and Claude's deep and serene Catholicism helped them survive this horror.

With Pierre and only a few others, I have the uncanny impression of carrying on a long conversation that keeps being interrupted, introducing new issues, and returning to old ones without ever losing an element of strong continuity. My fond hope is that it will continue until we are parted by death.

Pierre's parents ran an electrical supply store on the Quai. His father was made a war prisoner in 1940, but he got out and became a leader in the Resistance after we had left Tulle. So did Pierre, and one day after the war I asked him about the fate of the classmate whose sister was the prefecture employee who had deliberately displaced our file. The classmate had been loudly pro-German but seemed too weak to ever become a true villain. Pierre responded that he shared my feeling, and that in 1944 he made a point of arresting that classmate himself. What for? To tell him that he had been despicable but not a criminal. But if brought to court in the heat of the liberation, he would probably be imprisoned, then come back to live in his neighborhood and perpetuate the war hatreds forever. He set that classmate free, urging him to leave France for five years. The advice was followed, the malefactor's rage subsided over time, and he returned home as a neighbor one could live with. I was deeply impressed by Pierre.

Half a century later, in 1999, I was invited to address an exclusive scientific meeting in the Vatican and to bring Aliette. Surprised and delighted, we received an audience with Pope John Paul II. First we snaked through the theatrical private apartments, the inner sanctum, enormous rooms with little furniture, splendid paintings by Raphael, and Swiss guards dressed in uniforms designed by Michelangelo. Priests, bishops, and cardinals were all over, ranks being marked by subtle ribbons. All that pomp and circumstance was crying out to be reported to Pierre. Brushing up on the teachings of our French teacher, Mlle Tronchon, I found that my ability to write in eighteenth-century French had sadly decayed, but I did my best.

A few weeks later, Pierre responded with an unexpected tale. The countryside near Tulle to which he had retired had few inhabitants and a very few overwhelmed country priests, shuttling every Sunday between masses in all-but-deserted churches. Having become their resident lay spiritual leader, Pierre read them my letter, hoping they would be amused. Quite the contrary—they criticized me for consorting with His Holiness! Pierre defended me as having been merely a witness, not a coconspirator.

5

On to Lyon: Tighter Occupation and Self-Discovery, 1943–44

FROM 1940 TO 1942, life had gotten increasingly difficult. Still, we were not helpless refugees in a hostile land. In its narrow hollow, Tulle was out of the way and unnoticed when the Wehrmacht occupied southern France in 1942. The only nice hotel on the Quai became the seat of a rarely seen *Kommandatur*. A year went by with nothing worse than alerts that rushed us to safe houses. Léon finished high school. Then, one fateful day in the fall of 1943, our dear friend Monsieur Eyrolle dropped by nearly in tears with the news that his friend—that unsinkable politician Henri Queuille—had lost all influence and was himself threatened. That was the first time I heard who had been our anonymous protector in Tulle.

Now that we were on our own, life became much more dangerous. To keep body and soul together, most of my Jewish friends shared the risks by staying together. True to our antiherding instinct, our family decided it was best to split up: the boys on their own, and the parents on theirs.

Toolmaker in Périgueux Has a Very Close Call

Most fortunately—as happens every so often—an "Angel" appeared from nowhere, who Father could call on for help. On this occasion, we needed fake identification to get out of Tulle. Surely, this intermediary was chosen and paid for by some charity. But such help could not possibly have been given to everybody, and there was no way Father could have paid for it. The intermediary may have been the rabbi of Brive-la-Gaillarde, whom Father lobbied to help his elder son, whom he credited with every imaginable gift. Few aspects of that time give reason for regret, but I forgot the names of the charity and the Angel. I wonder who else received help.

As cover, Léon and I were advised to become apprentice toolmakers down the river from Tulle, in Périgueux. The shop consisted of one large ground-floor room with many old-fashioned machines, a few instructors, and a dozen or so young trainees. At night, we were assigned a room in a barracks close by, across from the rail station. Common sense dictated not talking to the other residents, who might include petty criminals and informers.

Not unlike sports, the bulk of training consisted of mastering a single but extremely arcane gesture. They gave us metal files, and from two pieces of scrap iron from a big drum we had to make a metal dovetail whose two parts would slide smoothly while allowing no light to shine through. When a part broke on a locomotive or wagon, a replacement could not be ordered from a warehouse, but instead had to be manufactured on the spot.

To teach this kind of repair, the files were very rough, and at first a perfect dovetail seemed impossible to achieve. But—not unlike cursive handwriting—it was "merely" a matter of mastering the absolutely precise muscle control needed to move a file back and forth in an exactly horizontal position. Having firm hands and a keen sense of space, I did very well, far better than most of the other trainees—who tended to drink heavily. I acquired a lot of self-confidence, and when I became a homeowner, my experience as an apprentice toolmaker in wartime was a boon.

Three of us—Léon and I and a mysterious fellow who came and went—stood out starkly. We did not in the least look or talk like apprentice toolmakers. But the manager knew our identities and was favorably disposed. Therefore, a routine set in for a while.

On my fake ID card, my birthplace was declared to be Bastia, a town in Corsica. The Allies had already landed there, so nobody could check and challenge me—it was hoped. But that ID card was not entirely foolproof, so I interrupted my routine and took the train to

Limoges, where a person I never met had volunteered to declare that I was living with her. The task of changing my ID card took more than one day, and the railroad station being always open and heated, I spent the night there, pretending to wait for a train. The station was otherwise empty, except for bums. One sat next to me and offered a drink from a filthy bottle. When I refused, he complained, *"T'es pas un pote"* (You ain't a pal). I was frozen with fear. In exchange for police tolerance, was he assisting them by reporting what he saw? My life was filled with potentially deadly situations that called for quick reckonings of odds. It was constant anxiety.

The next day, I visited the Second Police Precinct to pick up my "new and improved" identity card. No problem arose, but shortly afterward a person to whom I mentioned the event blanched and said, "Ain't you a lucky bastard. In the First Precinct, the boss is so-and-so, who was recently promoted from Tulle." Not a bad guy, but had I walked into his office, would he have arrested me? Fortunately, fate did not give him the choice.

One morning in November 1943, two men who identified themselves as policemen, followed by a police informer, entered the machine shop where Léon and I were pretending to be apprentice toolmakers. The town was abuzz with the previous night's big news: the occupying army's command post had been bombed, and the police were after the perpetrator(s). The visitors zeroed in on me and ordered me to put on my overcoat and beret and to show my identification—which was of course fake. The informer said, "No question. Here is your man," and then left with one policeman. The other stayed behind to reassure everybody not to worry. Léon and I thought this was our last day of freedom—or possibly of life.

My overcoat was distinctive. This prewar item, long misplaced and found by Father in a warehouse, had clear virtues—the cloth was not coarse like the substitutes manufactured during wartime and it was very warm, yet it required no coupons because it was not officially registered. The downside—and the reason it had remained unsold—was that it had a terribly loud "Scottish" pattern unlike any true clan.

Huddled together that evening after being visited by the police, with no one to consult, Léon and I wondered what to do next. The actual bomber must have been wearing a twin of my coat. The "good-cop, bad-cop" routine we had watched meant nothing, and the omens were grim. Flight was out of the question because we assumed that the police and some idle shelter neighbors were watching all the time. Long deliberation revealed no choice but to wait for the next opportunity to slip away unseen. In the meantime, we did not ask anybody about what was happening. We decided to act as if we were not concerned and stayed put with fear and clenched teeth.

We somehow managed to get hold of our Angel, and he arranged a more appropriate setup in the Lycée du Parc in Lyon.

Christmas at Saint-Junien

A suitable cover for the transfer came when our training center took everybody for a Christmas break to Saint-Junien. The poor Limousin countryside produced plenty of lambskin, and dating back to the Middle Ages, this spot has been devoted to glove making. Workers were extremely specialized, skilled, independent, and well organized—medieval guilds had morphed into single-minded and powerful unions with a strong anarchist tradition.

The street signs are all I remember of that place. In Nazi-occupied France late in 1943, one could find the boulevard Karl Marx, avenue Karl Liebknecht, allée Rosa Luxemburg, and similar homages to German Communist heroes. (Anarchists did not favor Lenin or Stalin.) Responding to my discreet surprise, a local explained that when an official foreign to Saint-Junien (for example, the prefect appointed by Vichy) was expected to visit, the signs were quickly replaced by the politically correct markers of the boulevard Marshal Pétain, avenue de Verdun, and the like. After the alert, the "rightful" signs were quickly put back.

Oradour-sur-Glane is a little town where the Waffen SS committed a horrible massacre in 1944, herding 642 villagers into a church and setting it on fire. It happens to be near Saint-Junien. So perhaps the SS Division Das Reich did not choose it at random but reacted to evidence of fierce local independence.

The Lycée du Parc

Much of the world was in turmoil, but at the Lycée du Parc in Lyon, where Léon and I boarded from January to May 1944, it was almost business as usual. I was greatly relieved, almost embarrassed. Actually, the normal lycée building—which I have not visited to this day—had become a military hospital, and the part I joined had been relocated from downtown to a bland building on a steep hill called the Croix-Rousse, a very proud working-class community long famed for its silk weavers and militant anarchists.

Practical aspects of life in hiding were not necessarily obvious. Our false ID cards were fine because providing them was a political gesture, an act of resistance. But ration cards were traded back and forth on a freewheeling black market, and inspection-proof ones carried a price tag that neither we nor our Angel could afford.

Therefore, the prime virtue of the Lycée du Parc in Lyon was that the business manager for live-in students agreed to turn a blind eye to our obviously "touched-up" ration cards. The fact that this lycée was arguably the best in the provinces was very fortunate for me, but an accident—unless our Angel's powers were truly supernatural.

To avoid lapses under pressure, my forged academic papers acknowledged my being from Tulle. Therefore, when—immediately upon arrival—I cast a chance glance into the humanities classroom and recognized a former Tulle classmate, I froze with terror. Looking him straight in the eye I said, slowly, "How nice to find you here. Do you remember me?" I gave him my assumed name. No response. I repeated the phrase once again. He smiled and answered, "What a surprise. Nice to see you. Of course I remember you." I breathed again—he would not tell on me.

My papers cautiously downgraded my baccalaureate from its dangerously conspicuous summa to an adequate magna. One day, a student approached. "I hear that you come from Tulle. You must have known Benoit Mandelbrot." "Of course, of course, I know him well." "Is it true that he is *un crack* who got a summa at the *bachot*?" Back in 1944, "crack" was French slang for a high achiever. Imagine my panic.

Did the student suspect the truth? Was he testing me? Trembling and with feigned nonchalance, I started telling stories about myself, how stressful it had been for "me," a mere future magna, to be in the same classroom as "that guy." I did not breathe freely until it became clear that the kid was simply curious.

My year and a half out of school was challenged. "Where have you been since high school?" "I was ill and followed the course at the École Universelle. It is a very good outfit." This was a down-market proprietary correspondence school that advertised heavily, but in the elite Lycée du Parc, nobody had direct experience with it, so they expressed surprise that such a school would be good. That answer bought time until the kids saw me perform, guessed the truth, and stopped asking.

Anxiety was rife. But even in the bleakest stage of the occupation, the worst horrors were never as systematic and uniform in France as in less blessed countries. They were largely localized as part of the bitter civil war between two historical sides of the country. Any one of my encounters might have led me to disaster, but none did.

Assigned to sit next to me in the classroom, Francis Netter was— like several other classmates—of an old French Jewish family and not in hiding. Their presence never created problems—a detail about Vichy France that invariably creates disbelief. Francis lived across the road from the school. Seeing that I was quite alone, his parents wanted to invite me for a meal, but I baffled and worried them— beyond an accent they thought was from Burgundy. They agonized for several weeks, fearing that kindness could backfire, then invited me, and Francis became—and remains—a good friend.

A Unique French Institution Nicknamed *Taupe*

The Lycée du Parc may be little known, but it is the keystone of the Mandarin system that seventeenth-century Jesuits had imported from China to France. That system's core consisted of more or less grand *grandes écoles*. To be admitted, one must pass certain "killer" exams. Having over time surpassed the *bac* in level of difficulty, those exams came to require intense preparation, and led to the development, parallel to the universities, of publicly supported *classes préparatoires*—cramming programs—that extend the larger lycées beyond the twelfth grade.

Taupe is the accepted nickname and *math spé* is the accepted abbreviation for the fourteenth grade—*mathématiques spéciales*. Similarly, *hypotaupe* and *math sup* (*mathématiques supérieures*) both stand for the thirteenth grade. Admission to those grades is largely based on performance on the *bac*, and the Lycée du Parc demanded at least a summa or magna. They let me skip the *hypotaupe* and enter at the midpoint of the *taupe*. For the last of the programmed four terms, I was a *taupin*, linguistically an extreme form of the American "nerd." In everyday French, *taupe* means "mole," presumably because continued overwork prevented us nerds from ever seeing the light of day.

These cramming programs are found in no other country, though Japan has something like them. All preparatory classes in France teach students to pass a very difficult test, therefore molding everyone to follow what was officially declared the straightest path to the best career. They are so tough that many good students (including some on the way to becoming great scientists) fail and repeat the *taupe* without any stigma.

A Geometer Meets the Love of His Mind

The few months that followed in Lyon were a transforming period of my life. Léon and I hardly left the school grounds. Even on Sunday afternoons, we rushed out after lunch and returned well before dinner—latecomers were denied dinner, and a meal elsewhere was well beyond our means. Moreover, we lived in deep fear of the German boss of the city: his name was Klaus Barbie.

Last but not least was a burning desire to catch up and do well. The war left no room for long-term ambition. Only the short term mattered. Bound to my desk, I worked at a rate that might not have been sustainable beyond these months, learning the ropes of the exams and honing my skills. Preparing for the exam by becoming a whiz at algebra was not a sensible goal. But, extraordinarily enough, events revealed a powerful gift of which I had been totally unaware.

During my first two weeks as a *taupin*, I wandered in a dark labyrinth, in a total blur—as expected. But in the third week, something remarkable happened, with no warning. Something so melodramatic that I can best express it with words from Puccini's opera *Manon Lescaut*—"*Sola, perduta, abbandonata in landa desolata*" (alone, lost, and abandoned in an unknown land). I too felt that way. Never mind that Manon was a Paris courtesan deported by the king to New Orleans. What matters is that utter despair was suddenly resolved— Manon's by the appearance of her lover, and mine by the manifestation of an unknown, powerful force.

Our math professor, Monsieur Coissard, had just joined the Lycée du Parc, where he was to spend a long and admired career. Even within the elite group of *taupe* professors, he was outstanding. About half of every day was spent with him, and he would go to the blackboard and describe a very long problem that—building upon generations of educators—he had deliberately contrived to require absurdly complex calculations. The problem was invariably stated in terms of algebra or analytic geometry.

My inner voice was restating the same problem in geometric terms. During all that time in Tulle, I had relied on those outdated math books filled with many more pictures and fuller explanations and motivations than the books of the 1930s—or of today. Learning mathematics from such books made me intimately familiar with a large zoo, collected over centuries, of very specialized shapes of every kind. I could recognize them instantly, even when they were dressed up in an analytic garb that was "foreign" to me and, I thought, to their basic nature.

I always started with a quick drawing, which I soon felt lacked something, and was aesthetically incomplete. It would, for example, improve if transformed by operations called simple projection or inversion with respect to some circle. After a few transformations of this sort, almost every shape became more harmonious. The ancient Greeks would have called the new shape "symmetric," and in no time searching for and studying symmetry became central to my work. This playful activity transformed impossibly difficult problems into simple ones. The needed algebra could always be filled in later. Hopelessly complicated problems of integral calculus could be "reduced" to familiar shapes that made them easy to resolve. I would raise my hand and describe my findings: "Monsieur, I see an obvious geometric solution." I quickly grasped the most abstract problem that the teacher could contrive. And then—with no effort, conscious search, or delay—I continued along a path that somehow avoided every difficulty. As the term progressed during that winter in Lyon in 1944, my freakish gift was revealed as strong and reliable.

In a way, I was learning to cheat. But my strange performance never broke any written rule. Everyone else was training for speed and accuracy in arcane but teachable arts—algebra and the reduction of complicated integrals. I managed to be examined on the basis of speed and good taste in, first, translating algebra back into geometry, and then thinking in terms of geometric shapes. My analytic skills remained so-so, but that did not matter—the hard work was done by geometry, and it sufficed to fill in short calculations that even I could manage.

My Lyon classmates keep in touch with one another and with me to some extent. Recently, Francis Netter wrote me that I had been the best *supériorité absolue* in mathematics. It was not my high aptitude for the sciences that surprised him the most, but the broader learning I possessed. One day, he recalled, we wandered around together, and I described enthusiastically a work and an author he had not previously heard of: *Buddenbrooks* by Thomas Mann. Where, when, and how did that book of Mann's come into my hands? Did I praise Mann because he opposed (and fled) Hitler?

In 1973, I visited M. Coissard near Chamonix, in the French Alps. I met his wife and his successor in Lyon, who vacationed nearby. A truly emotional reunion! Coissard told his side of the story—how deeply, during that winter of 1944, my ways had affected his own life and that of his father, a retired *taupe* teacher himself, who lived with him. Both spent long evenings and weekends looking within the exams for old or new problems that I could not instantly "geometrize." They never succeeded in stumping me.

Where did that gift come from? One cannot unscramble nature from nurture, but there are clues. Szolem lived a double life as a week-day mathematician and a Sunday painter, and his son is a physicist and a painter. I mix mathematics and art every day. My gift for shape may have been saved by all the complications that marked my education during early childhood and the war. Learning to be fluent at manipu-lating formulas might have harmed this gift. And the absence of regu-lar schooling in art may have influenced many life choices, ending up not as a handicap but as a blessing.

Oddly, the *taupe* curriculum included freehand drawing. Before photography, engineers were supposed to illustrate their own work. Most students were all thumbs, but that family gene made my work extremely precise. Our subjects were mainly overused plaster casts of famous sculptures from the Louvre: the Venus de Milo (smooth and easy), the Victory of Samothrace (hard-to-draw wings), or the head of Voltaire by Houdon (a most challenging wig). The drawing master collected our efforts and returned them with grades and comments. The school had forgotten to announce the new student, so when the master exhibited my first drawing, he commented: "It seems that this is a practical joke—a drawing by a virtual student from the outside. I would love to see the students in the arts program do as well." I stood up and introduced myself.

★ ★ ★

The mental and physical stress of the *taupe* was immense, but I managed. The effect that time in Lyon had on my life has been extraordinarily deep and durable.

6

Horse Groom near
Pommiers-en-Forez, 1944

AFTER THE ALLIES' JUNE 1944 LANDING in Normandy, the Lycée du Parc rushed to close. Everyone was chased from the dormitories and urged to leave Lyon. Our Angel appeared! Léon and I were told to report to an office in Roanne, a midsize town west of Lyon, to be assigned to neighborhood farms while we awaited further instructions.

Léon's farm made him work strenuously—and he did. Mine did not go as well. It was near a small town, Saint-André-des-Eaux, in a formerly volcanic region not far from Vichy. Its sparkling water was iron-heavy and covered the taps and all the landmarks with a smooth rust-colored patina. To work in mulch and manure, we wore wooden shoes—plain chunks of wood—on naked feet, and a kind of leather grew on a new wearer's skin after the scabs came off. One day the yoke of the oxcart fell on my knee, and for several days I could hardly move. My boss, a kind old farmer, told me that he was better off without my help. I agreed.

The Roanne office next sent me to an isolated horse farm, directing me to take a bus to Saint-Germain-Laval and from there to walk east, beyond Pommiers-en-Forez, through a fertile agricultural region between Roanne and Saint-Étienne, then a major center of mining and steel.

I reached Le Châtelard late in the evening and was greeted by a generously built woman who the next day was revealed to be Countess Suzanne de Chansiergues d'Ornano. She had inherited Le Châtelard from her mother and lived there with her husband—the count—and her father, Monsieur de Rivière, who may have been sixty but suffered from arthritis and to me looked ancient. There were also a few house servants.

At lunch the first day, I was told that they were raising horses—animals I had not dealt with since my summer in Belarus. At one point in the conversation, M. de Rivière became animated and began rambling, "In 1913, my horse Phoebus won the Derby de Lyon. He was a trotter." He then recited the horse's pedigree back several generations. No one at the table paid any attention. I learned that the horses in residence were Anglo-Norman half-breeds—a subtle balance between the extraordinary beauty (but notorious fragility) of English Thoroughbreds and an ability to perform marketable work. The art of breeding these horses largely consisted of hiring one of the stallions—ranging from pure Thoroughbred down to mixed breeds—available in the Haras Nationaux, a stud farm maintained by the government. Le Châtelard had dwindled at that time to two breeding mares, Rêveuse and Respectueuse, chestnuts with black mane and tail, and their foals.

That night at dinner, M. de Rivière again became animated: "In 1913, my horse Phoebus won the . . ." I interrupted and recited the horse's pedigree without one mistake. "*Ça, par exemple!* Nobody ever listens to me, but you did. And you remembered everything. You can't be altogether bad."

Shortly afterward, M. de Rivière confessed that he needed a groom for his horses, and with everybody away at war, the choice was between me and "Jules" (his actual name escapes me now). "Jules knows everything about horses," he said, "and you know nothing. But he is a thief, and you look honest. I take you, and you can continue to eat at the master's table with us." *Phoebus* is the ancient Greek word for the sun. Long after his death, that horse brought the sun to shine on me. His name is one I shall never forget.

As a premium, I got a glimpse into a world of country gentry that would otherwise have remained completely closed to me. When M. de Rivière was young, his prowess as a horseman had won the love, the hand, and the roof of the heiress of Le Châtelard, who was no longer living. His closest friend was the daughter of his wet-nurse when he was a baby. A widow, she came on visits from nearby Saint-Étienne. The estate also included two little-used farms.

Madelon, an ageless mixed-breed draft mare, did the heavy work and was unremarkable. Poor Rêveuse and Respectueuse were idle and skinny, but I learned to make them look good enough. Union Sacrée, an old half-breed, had long been retired. Every Sunday, as prescribed by my boss, I took her for a bit of exercise that was so strictly defined that I swear she knew it by heart. I was taught how to do her toilette with a checkerboard backside, to hitch her to the carriage, to hold the reins properly when bringing the countess and her father to church, to wait with her in a certain tree's shadow, to bring her back home with her load, and finally to take her to her stable. I asked M. de Rivière, "What will happen when she becomes sick or helpless?" "I watched that horse being born, and I will never let her suffer. I shall take her behind the stable and shoot her with my old handgun—right between her eyes."

M. de Rivière decided to bring his mares and foals to a competition held on the racetrack near Feurs, the tiny historical capital of the Forez. I was told that at one time the cavalry used (or even sponsored) this competition to select new horses to buy from the breeders. By 1944, however, the word "cavalry" had shifted to denote tanks, and that competition had withered to an occasion for old friends or foes, isolated in their estates, to get together for drinks and gossip. Horses smell, especially when stressed and hot in summer, and so does their manure. Many horses brought together on a muddy field emit a stench—and neighing noises—that I still remember as I write.

The cavalry's criteria put Rêveuse and Respectueuse behind all the other mares—immediately preceded by one owned by Jules. His bay was broader and more muscular than our chestnuts—thanks to the oats he was reportedly stealing from us. One of ours limped, but the other was not that bad. M. de Rivière talked to his friends, and Jules's mare was bumped to second from last, while our "good" one was promoted. Madelon took us all home, with the reins in my hands.

The time arrived to give names to the foals. The initial letter of a registered horse's name was set by the studbook and was cycled around. Rêveuse and Respectueuse, both born in 1940, shared an initial *R*. The compulsory initial letter in 1944 was *A*. Jules called his foal Algérie, and M. de Rivière—ever faithful to Greek names, which reminded him of his beloved Phoebus—settled after days of seclusion on Aphrodite and Apollo.

Sad to say, Apollo had a limp—like Aphrodite's mother—and had to be sold. The decision was made just in time for the Foire de la Bautresse, an international fair held near a place called Boen every year since the fourteenth century, in peace or war, in prosperity or famine. Madelon, Apollo and his mother, M. de Rivière, and I set forth to the fairgrounds to sell Apollo.

Madelon was left near the carriage in a parking lot, and we moved to a spot reserved in advance. I was told to stand by my horses and tell all dealers that the price for the colt was forty thousand francs. M. de Rivière then left to join his pals. The horse dealers of half of Europe (or so it seemed) soon descended on me and immediately recognized a city rube. "Forty grand for both—that is a fair price." "No, no. Forty for the colt, without the mare." I don't recall any second look at the limping little beast. Every so often, M. de Rivière came back to check on offers. Nothing to report.

Dusk came and we trudged back home, a disappointed and exhausted little caravan moving so very slowly. Along the way, a noisy carriage filled with peasants drinking and singing passed us, and trotting behind was a colt with brown-yellow hair. A passenger hailed us. "So you did not sell your foal. I bought myself this nice one for twenty-two thousand francs." Off they trotted, with M. de Rivière muttering, "No breeding at all." At that point, the peasant holding the reins slowed down to let us catch up, and another passenger shouted, "I can use a horse at my farm. I shall pay you twenty thousand francs if you bring it to my place." He also shouted his name and address, and then their carriage sped up and passed us again.

A limping Anglo-Norman Thoroughbred had no value, yet consumed expensive food every day. M. de Rivière had to sell, even though Apollo had not yet been weaned. We took his mother along with a very long leash. "What for?" I asked. "Wait and you will see."

We reached the farm, and twenty thousand francs in small bills changed hands. After we had our drinks, M. de Rivière remained seated, and the farmer became impatient. "The horse has been paid for, and I have tons of work to do. Good-bye." "But you have not yet bought the bridle from the groom." "Is the groom this fellow?" the farmer asked, pointing to me. "Yes. Let me explain. The bridle itself is a worthless old piece of leather. But he had to work hard, and paying him for the bridle is a nice way of giving him a tip." "No way, no way, nobody heard of such nonsense! Good-bye." As we left, Mr. de Rivière muttered that he would "buy" the bridle from me, but I assured him that nothing was expected.

The ride home was slow and exhausting. The mare dreadfully missed her foal and ran here and there seeking him. That is when the very long leash came in handy. I gave her rope to run, then pulled her back when feasible. That night, M. de Rivière had to stay up to milk and soothe her; I could not help.

★ ★ ★

Politically, the horse-raising gentry of Forez were far from raging radicals. At lunch and dinner, we listened in respectful silence to the obviously biased news on Vichy radio. However, their resigned expectation of German victory and their acceptance of Marshal Pétain eventually wavered. I cajoled them, first to listen to Swiss radio in French, then to France Libre in London.

The result was unexpected. Hearing London radio describe General Charles de Gaulle's background, they all perked up and recognized him as one of their own—while I was becoming somewhat dubious. By pure fluke, through a flimsy partition, I had heard his famed June 18, 1940, appeal on the neighbor's radio, urging the French nation to continue fighting. But during the war, neither side had found it politically advantageous to inform the French that this general had long been close to Pétain—he was, in fact, the godfather of his son Philippe. Only after the war did some writers argue that Pétain and de Gaulle were secretly playing the two sides of a "foreign" conflict. Incidentally, the patronymic de Gaulle is not aristocratic; in Flemish (the tongue of northern Belgium and France), it means "the horse."

My wartime hosts in Forez had wondered who I really was, never quite figured it out, but always behaved in a thoroughly civilized fashion. Perhaps, in their isolation, my unending stories had been entertaining. Many years later, Aliette and I drove by. France was becoming very rich, and Anglo-Norman horses were all the rage. But that estate looked abandoned, plain, and charmless.

One final irony—our horses were either not broken in or too old, so I never learned to ride.

7

Alleluiah! The War Moves Away and a New Life Beckons

ON AUGUST 15, 1944, the Allies landed in southern France. Shortly after, the occupying forces on the southern front broke position and rushed farther north, skipping the Forez altogether. On the northern front, Paris was liberated on August 25. Liberation was an explosion of joy combined with a settling of accounts.

The war, with its fears and deprivations, left a mark on me that would never wear away. That mark persists in the obvious big things that have shaped my life. It also persists in small things—I still can't throw away paper that might someday find a new use.

* * *

My luck in wartime had been stark and simple. After just barely escaping the coming horrors in Poland, I had managed to survive the occupation of France, with its periods of astonishing "normality" alternating with hair-raising episodes. Not only was I never trapped, but—for one reason or another—I was repeatedly given a pass and never denounced. I received enormous help, and more help must have come along without my knowledge.

I had absorbed enough *taupe* to do very well on the exams—but not nearly enough to fit any of the stereotypes that matter in France. Even more than my years in Poland, wartime made me "different" for life.

My weeks at Le Châtelard put me in excellent physical shape. Yet, for years to come, I was continually told that I looked older than I was. This changed only after I met and married Aliette.

* * *

As soon as it was feasible, Léon and I rushed to reunite in Roanne. Amazingly, a train ran west from there to Clermont-Ferrand—on schedule! Apparently, in those poor highlands, the railroad bridges had not been worth destroying. But the connecting westbound train had already left Clermont-Ferrand. At that time, no central authority set the clocks, and the southwest set its one hour ahead of the southeast.

When we arrived at Tulle, our foreboding melted. Marvelous, incredible surprise—Mother and Father were waiting for us at the station! For quite some time, they had been meeting every train from the east and returning home empty-hearted.

Deliriously happy family reunion with no one missing! We soon learned, however, that—like Oradour-sur-Glane (near Saint-Junien)—Tulle had witnessed a major abomination, which the news reports we heard had failed to mention. After we had left, the Resistance became organized and very active. In response, not long before being forced to move north, the occupying forces that smashed Oradour also hanged nearly a hundred young people from lampposts and balconies of a square right next to where we lived. One victim had been a classmate. Both of his parents were schoolteachers. Nice, brilliant, and all too secure, he was often heard expounding in a loud voice classic left-wing ideas that I would not have dared even whisper. Was he picked at random or fingered as a troublemaker?

The bold plan our parents had devised—bless their hard-won survivor skills—had let them and their sons cope with events separately. This bet, the riskiest of our complicated lives, worked better than any realist could have hoped. Parents and sons soon returned separately to Paris. The railroad bridges across the Loire River near Orléans had been bombed out; passengers lugged their bags along a pontoon bridge. But this was nothing.

★ ★ ★

To this day, my spoken French preserves traces of slum Parisian and Limousin. On balance, in my heart—though the Tulle where I lived has been swallowed by history—I shall always remain a Tulliste. I visit as often as I can. It becomes ever harder to associate what I see today with the tiny, ancient farms, the empty villages with big monuments that list the dead of 1914 to 1918, and the 1944 torrent of blood.

An ever-optimistic Pangloss would say that flight from Poland and survival in wartime France were proper preparation for a life that never stopped being overly interesting.

★ ★ ★

Nearing the age of twenty, I was acutely aware of entering a second stage of life and intensely hoped it would not be so hard. But the past cannot simply be left behind, especially a past like mine, so sharply synchronized with the Depression and the war. I never had any leisure time to "find myself"—except for my wild mathematical dream. As a student, I would do well, but being in control of my life was an unfamiliar situation. During this second, twelve-year stage of my life, I was not going to manage elegantly—as will be seen. So, in time, I deliberately provoked a belated third stage.

Part Two

My Long and Meandering Education in Science and in Life

*"Maverick" is a term I wear with pride. I lived as one and intend
to die as one. I slid into being a maverick starting at the moment
of my coming-of-age. A youthful decision set me on a maverick's
lonely ride. Its consequences took a long time to develop.
Had I died or retired in midlife there is no way they
could have been understood.*

My Keplerian dream begins to unfold.

8

Paris: Exam Hell, Agony of Choice, and One Day at the École Normale Supérieure, 1944–45

By September 1944, most Parisians, including my family, had moved on mentally to new and different challenges. Even before the actual armistice on May 8, 1945, the end of the occupation was an infinitely welcome relief, but it also posed a complicated turn in my life.

My next task was to seek the best university that would both accept me and encourage or at least tolerate two self-imposed goals. I wanted to keep close to geometry and to prepare myself to realize in some way that Keplerian dream I had formulated not too long before. The scary exams proved a cinch and brought about the first, the freest, and most agonizing professional choice of my life.

The Holy Grail: *Les Grandes Écoles*

The École Normale Supérieure and the École Polytechnique drew applicants from the whole of the country and were absolutely the best France could offer in the sciences. They used to nearly face each other on the south and northeast sides, respectively, of the Panthéon in the Latin Quarter.

In an extremely rough way, the École Normale Supérieure—or Normale, rue d'Ulm, or ENS—was a miniature Cambridge or Harvard, casually nicknamed Gnouf. Of about two hundred rigorously screened candidates in math and physics, it was entitled to accept twenty-five for the 1944–45 academic year. The war had affected many, and that year's admissions committee ended up accepting only fifteen. The school's name suggested that it trained male teachers for the elite secondary and higher education. (The popular primary track described earlier recruited from the numerous *écoles normales d'instituteurs.*) By remaining tiny, it grew increasingly prestigious and evolved to train researchers and teachers for the universities and classes like the *taupe* I attended in Lyon. In 1945, its reputation depended on how a field was doing—high in pure mathematics but not in physics.

The École Polytechnique is far smaller than MIT. It is generally called X, but the name I prefer is the one the students and alumni used in my time: Carva. This is an abbreviation of *boîte à Carva,* after the long-serving dean Moïse Emmanuel Carvallo, an activist and half-mythical figure. Since my time, the school has become coeducational, and its enrollment, curriculum, and opportunities for graduates have continually broadened.

As I knew since the time Father introduced me to Paris, Carva was located at 5, rue Descartes. From two or three thousand top-level candidates, its entrance examinations selected about two hundred. Compared to Normale, Carva could be called either more diversified or weakly focused. Everyone knew that its alumni ran the gamut of French life—they could be found in many agencies of the state, as well as in private banks and businesses. A few were priests, monks, professional writers or musicians, even local or national politicians. Its long role as a military academy (it was the model for West Point) had largely disappeared. Early in its glorious history, it had produced the bulk of French scientists. This was followed by a lengthy gap, but the tradition has since been revived.

How to match eager candidates and limited-enrollment schools? Before the Revolution, well-paying offices were inherited, granted by sovereign pleasure, or purchased. By contrast, the *grandes écoles* recruit on merit. Their entrance exams were (and remain) the French counterpart of the cruel-by-design rites of passage practiced among the "savages." To prepare for them, there were cramming programs like the one I took in Lyon at the Lycée du Parc.

In the fall of 1944, back in Paris, I transferred as a resident student to the Lycée Louis-le-Grand—the crème de la crème, named by King

Louis XIV himself. I sat in the class of Monsieur Pons, hardly ever speaking to him, but cramming by myself. The delayed exams began in December 1944 with a week of written Normale and one of written Carva, and ended in January 1945 with a week of oral Normale and one of oral Carva.

At Normale, one mathematics test was so long the proctors called a brief break and fortified each candidate with a bowl of hot broth. Later, texts in several different languages were handed out, and we had to translate any two. To English, an obvious choice, I added Latin!

By design, both exams were extremely difficult, sufficiently so to ensure that, typically, only the top man managed an average above 16/20. Rumor had it that, as of 1945, the all-time record had been set around 1885 by Jacques Hadamard, Grandfather's senior guest at that 1930 dinner in Warsaw and later my grandfather of the mind.

Unexpected Triumphs at Normale and Carva

In January 1945, the week between the written and oral exams, I was racing across the Latin Quarter when my mathematics teacher, M. Pons, hailed me in the street, and we had our first and last private conversation. "Let's talk about the big math problem at Polytechnique. I could not solve it in the time allowed, but examiners say that—in the whole of France—one student did solve it, and he is from my class. Could it be you?" "Well, I did solve the entire problem—including every optional question at the end." "How did you manage? No human could resolve that triple integral in the time allowed!" "I saw that it is the volume of the sphere. But you must first change the given coordinates to the strange but intrinsic coordinates I thought the underlying geometry suggested." "Oh!" And he walked away, repeating, "But of course, of course, of course!"

When the exam ordeal ended, my grade was 19.75/20. Nobody ever received 20/20—ever! For this and other top mathematics marks, rumor appointed me the best math student in the country that year. Everyone seemed to know of my skimpy formal preparation, so I was credited with a feat that would be remembered for years to come.

I was supposed to take those exams as practice for a serious second try. But that 19.75/20 was approached by some other very high marks, mostly in the additional math exams. Also, I wrote very good French and reasonable English and had high grades in freehand drawing. Somehow, subpar grades in "lesser" tests did not register, and I was widely believed to be number one. A major moment in my life!

Plain and simple, not only had I survived the war, but in France *I had it made for life.* Of course, nothing could guarantee that I would mature into a great scientist—or a great anything. But either school could open every door and provided a kind of automatic lifelong insurance. All this was simply beyond belief. Only nine years since my move to France, only months since the liberation, and still officially residing in that slum of Belleville, I was in no way ready for such choices.

★ ★ ★

As I look back, it seems to me that great opportunity in one way was wasted but in another was used in the best possible manner. For thirteen years, my suddenly acquired "capital" was not wisely invested and was ostensibly squandered over a period of slow maturation and wandering. Then I moved to the United States—where French credits had no value. However, there I managed to design a career to match my skills and tastes—one that lost out on all the French benefits earned by those exams but perfectly fit the dream I had conceived during the war.

Reunion with Uncle Szolem

One day when I was returning from a Normale exam to my quarters at the Lycée Louis-le-Grand, a man hailed me in the lobby. For a moment, I mistook him for Father. But he was younger and free from Father's stigmata of perpetual deep worry. Sure enough, it was Szolem—back in Paris from the war years spent at Rice in Houston and then with the military in London.

He noted I looked fit, rather than starved, and I told him why. We worried about those in Poland but reassured each other that in his own family and mine everybody was alive and well. His wife and son were about to return, and I put him in touch with Father.

Then we moved on to the exams. "How far along are you now?" "I just took the big written math exam of Normale." "How far did you get?" "All the way to the last question, and I can't think of any bad mistake." "Splendid, congratulations, splendid, splendid. You are a shoo-in. I am so glad. You are lucky. You will go to Normale and experience something marvelous that all my friends went through but I missed."

For many days, months, and years to follow, Szolem told me about the inner workings of the worlds of mathematics and science, which he knew well. He spoke often about his mentor, Hadamard, and his contemporaries. He described the colorful André Weil and the completely fictitious but increasingly influential group of Young Turks that Weil conceived, organized, led, and named Nicolas Bourbaki.

All this was fascinating to me. Much was extremely attractive, but Weil and Bourbaki were positively repellent. Right after the war, I was wary of secret groups and charismatic leaders, and this leader's taste was extremely far from mine. It will be seen shortly how this affected my life. Combining idealism and practicality, Szolem described very frankly both the system's greatness and its warts—such as the pervasive patronage and the widespread inbreeding and nepotism facilitated by the fact that, even in mathematics, judgments of value are subjective.

Family "War Council"

It is my impression that among my fellow students who did well enough to have a choice between Normale and Polytechnique, very few agonized. Regular schooling identifies sensible ambitions, and my classmates had been preparing over much of their lives. By contrast, I was both underschooled and suddenly overadvised. Only months before, I had been desperately focused on staying alive. Now a marvelous long-term choice became available for me alone to make.

A detail that became very important: entering Normale, students chose between mathematics or physics but could easily switch; Carva, to the contrary, allowed minimal advance planning, and opportunities were tightly restricted by the rank a person received at graduation.

The high stakes terrified us all, and my parents did not trust my teachers. So a family "war council" was called to help: Szolem and a second cousin and close friend—the leading physical chemist, Michel Magat—met in February 1945 in our Belleville tenement.

Uncle and Cousin were brilliant and forceful, politically engaged but unbelievably partisan and naïve, as it soon turned out. They battled against each other and Father for my future and my soul. Exact words are, of course, forgotten, but the message remains clear in my mind.

UNCLE: Carva transforms bright students into soulless bureaucrats who can't run anything properly. They won World War I, but lost World War II. Follow the path I took, and add one thing I missed. Go to Normale. No career brings the rewards of pure science. It gives you both freedom and insurance, because the alumni take care of their own. If you are unlucky and discover nothing important—but don't worry, you will have no problems—you will become a high school teacher. No career comes closer in serving society, and you will be happy and proud of yourself.

COUSIN: Inescapable social and political forces are about to abolish both schools. Carva is a bastion of obsolete ideas and old ways. They will teach you nothing, only make you feel you belong to the elite. Normale is just as bad. Consider the École Supérieure de Physique et de Chimie. It is supported by a city— not the state—and knows how to train people to become down-to-earth scientists.

FATHER: Don't listen to either of them. Thriving as a scientist is a lottery. Szolem won a jackpot by being smart but also by coming to France at precisely the right time. But France, Europe, and much of the world are a total mess—no one can predict what will happen next. Cousin's predictions are not serious. Besides, if the Russians help the Communists come to power here, you may be forced to pull up roots once again and move to a new country—Brazil, Argentina . . . who knows? Since we married, Mother and I were wiped out six times by events over which we had no control. Also, never forget something basic: professors are civil servants. Trouble may leave you somewhere—as it did Mother—with a worthless foreign certification. Keep away from state-certified fields and large national organizations. Education, health, and law are the plague. Go for broad engineering skills that every country will need under every political regime.

Father, a skilled survivor, deeply admired scholarship and practiced it—but only to the extent that circumstances would allow. He strongly believed that a scholar's happiness and independence hinged on a steady income largely independent of uncontrollable events. This attitude was forged by the chaos he had experienced. One hears the same advice today all over the media: don't count on lifetime protection from one employer. Many years before, Father had given that very same advice to his twenty-year-old brother, Szolem.

Years later, I realized that Father's thinking had a far broader perspective. He was quite impressed by the work and misfortunes of a Portuguese Jewish philosopher born in Amsterdam, Benedict Spinoza (1632–77). Spinoza's community shunned him, yet being a skilled lens grinder in tolerant Holland allowed him to think freely. His spiritual power stood in sharp contrast to his political powerlessness. In our family, achieving political power crossed no one's mind.

Similar family fights have occurred in the lives of two scientists I came to know. To help the biologist Jacques Monod decide between biology and music, his influential father appointed a committee. It reported that as a biologist he would match Pasteur and as a musician he would match Mozart. He chose biology and won a Nobel Prize.

More important for me was the great mathematician John von Neumann, to be introduced later. Around 1920, Hungary, his motherland, was under a cloud of uncertainty far worse than Poland in 1920 and France in 1945. His rich father wanted him to play it safe and study chemical engineering, but agreed to hire a young Budapest professor named Michael Fekete to determine whether "Janos" should also be allowed to seek a Ph.D. in mathematics. The advice was that he should do both. He perfected an alloy whose composition is not expected to ever be encountered again.

Much in my life is easily traced back to that family war council. In effect—in a most fruitful draw—Father and Uncle both won and earned my everlasting gratitude. Their respective influences did not just mix in my life—they simmered slowly under the blows and the heat of successive trials and errors, eventually yielding something quite distinct from each of them—a new alloy.

A Die Cast One Day Is Retrieved the Next

At first, Uncle's academic position and personal authority prevailed, and I registered at Normale in extremely high spirits. I had every right to be proud of myself. I had survived the war, thanks mostly to help and luck, but also to fast thinking on my feet. Then—my acrobatic feat—I took this exam almost cold and came out near the top.

On my first day at Normale, the deputy director for the sciences talked to me in the threshold of his office. We discussed my formal status as a foreign citizen who had passed the regular ENS exam and hoped to be naturalized. "There is no difficulty whatsoever," he assured me. "As soon as your naturalization comes through, you will become a regular student. Till then, you will have to pay tuition and board. Your situation is rare but not unique." One precedent he managed to recall was a philosopher then at the height of his fame, Henri Bergson (1859–1941), to whom—as he observed—"this initial complication did no harm." We agreed that the precedent was flattering and promising.

Unfortunately, as the first day went on, a good look around made me feel dreadful. "What am I doing here? This is absolutely the wrong place for me." I finally faced a reality that Szolem had described to me—a reality I had previously disregarded. The Bourbaki cult was becoming dominant in pure mathematics, and Normale was about to be taken over. It was indeed the absolute worst place for a strong-willed person with already clearly defined tastes. I spent the day agonizing, could not imagine a good reason to stay, and went back home for the night.

By the next day, I had yielded to Father, and returned to Normale to resign. Léon often reminisced about everybody's surprise at my sudden change of mind. This key decision to switch schools— although it complicated the second stage of my life as a scientist— proved to be the right one and dominated my whole career.

The decision was widely misunderstood and criticized, and some potential friends never forgave me. Szolem became upset and afraid, the way any fanatic, scientific purist fears new alloys. Even now, it is insinuated that I did something very wrong.

The Weather and the Mood of the Day

Individual decisions are randomly influenced by history in the making. In prosperous and happy times, the influence is very gradual, but not so on that day in the war-weary France of early 1945. The family war council was inevitably affected by the historical "microclimate." The class of 1944 made choices in the middle of the abominable last winter of the war. How could this fail to matter? Only weeks had passed since an enemy counteroffensive in the Ardennes near Luxembourg created the scary Bastogne Bulge and threatened to push back the war's end. Physically, Paris was nearly intact—but cold, bleak, and desolate, reeking of poverty and decay.

Had I been a true believer in French mathematics à la André Weil, none of this would have been noticed. But I was not, and the mood of the day inevitably affected my decision. Sunny weather, good progress in the war, and a buoyant political situation might have made dwelling in the lay monastery of Normale acceptable. I shudder at the thought.

Intergenerational Conflict Among Immigrants

Around March 1945, Szolem resumed his chair at the Collège de France. At his first lecture, I was the only young person present, and he kept it at a level I could follow. The attendees proceeded to the cobblestone courtyard, mostly to exchange news of who had or had not survived the war.

I recall clearly Szolem introducing me around and commenting on my scandalous choice in a tone appropriate for a funeral: "Having entered Normale, this boy has left on his second day and is about to enter Polytechnique." He could not understand why anyone would look for mathematics different from his or that of Bourbaki.

Michel Loève (1907–79), a Jew from Alexandria, Egypt, was there and spoke reassuringly: Polytechnique was of course second best, but fine, since I would study under Paul Lévy. That moment was my introduction to a great man and a major figure in the exciting field of probability theory. This encounter with Loève earned my gratitude. In due time, it would combine with other forces to steer my Keplerian dream toward the theory of chance.

While Szolem had been anything but bland in his twenties, age and success had mellowed him. He was liberal on most issues—except those close to his heart. Since I wouldn't follow in his footsteps, we had terrible fights. Until I pinned down what exactly I wanted to do, I kept losing—of course. He never understood my aspirations, continued to worry about my very bad taste and its inevitably horrible consequences, and felt to the end of his life that my intellectual gifts had been squandered.

A major dynamic of our relationship was simply a classic reaction against a powerful father figure—in my case, not Father, but Szolem, twenty-five years older than I. This involved a theme favored in fiction and history: intergenerational conflict among immigrants. In our family, fleeing Poland put Szolem in the first generation; I stood on his shoulders and belonged to a freer second generation.

Similar consequences result when a law subjecting a population to stringent restrictions is suddenly overturned. Their natural reaction is to keep complying. Szolem's youthful fling on the political and literary scene contradicts, but was transient. On the more important scene of mathematics, Szolem fit the first-generation stereotype by acting as a prudent conformist who promptly joined the soon-to-be-powerful Bourbaki.

To the contrary, I fit to a tee the second-generation stereotype, which today's France knows best through the children of immigrants from Africa. I never turned to political violence, yet became a non-conformist, a permanent questioner who managed to thrive without either joining an existing school or creating one for the few formal students I had. Therefore, seen from a distance, the path of Szolem's scientific life seems straight as an arrow, while mine was . . . unquestionably fractal. But maturity brought out many similarities. It became important that we were both "ideological refugees" from utter abstraction. Sierpiński intellectual and political views made Uncle flee Poland, and Bourbaki made me leave Normale in 1945—and France in 1958.

Two examples of sweet irony: Szolem loved and faithfully served through his life two topics of truly classical mathematics: the Taylor and the Fourier series. In the twentieth century, both developed into fields self-described as "fine" or "hard" mathematical analysis. They forgot their roots in physics, except for a massive contribution from another man who was to play an important role in my life, Norbert Wiener.

After Szolem made me learn these topics, I flew away—but never jettisoned what I had learned. In Szolem's theorems, the list of assumptions could take pages. The distinctions he enjoyed were elusive, and at his preferred level of complexity, no condition was both necessary and sufficient. The issues he tackled had a long pedigree within pure mathematics. This was for him a source of pride, but was for the younger me a source of aversion.

A wandering scientist should never say never, and history shows that beautiful parts of abstract mathematics can well slumber for a while and become disconnected from their roots in reality. They may seem to be stone dead—yet should never be called exhausted.

A second example concerns the intellectual landscape previously visited by Szolem and his students and friends. In due time, a persistent scientific study of roughness led me to encounter increasing depths of wild complexity—and therefore to cease to expect the world to be fundamentally mild and simple. To my initial astonishment and ultimate delight, I encounter again and again the hard messiness found in Szolem's mathematics. Its practical applicability revealed that it reflects the irreducible messiness of where I have chosen to work—the scientific frontier.

An Unexpected and Much-Needed Pause

The last-minute switch from Normale to Carva was made possible in January 1945 because of a wartime quirk. Normale had vacant dormitory space, but Carva did not. So, from February to August 1945, my class waited for space to become available. The army drafted my future classmates into a special unit. I volunteered to join them but, being a foreigner, was turned down. Thus, for half a year, my schooling was interrupted once again—by odd jobs.

Many people I know and respect value efficient processing of youths and view "wasted time" as harmful, even threatening, or immoral. I had no choice. Moreover, I think it helped me grow up—a valuable gift from fate. Much later on, I was happy when both my sons had reason to take years off.

Let me elaborate. Good wine or cheese must not be rushed. So why rush good humans by pressing a cookie cutter on a malleable young mind? As that stuff sets in, it preserves for life the shape of the mold. Today, taking time off is tolerated, but still not in the hard sciences. Among my old classmates, many acknowledge lifelong distress at having never been given a break.

Sirens Hawking All Kinds of New Propaganda

I was waiting for a top school to be ready, but money and food were a constant issue. I did not mind eating quite often at a soup kitchen in Belleville, probably supported by some American Jewish charity. Most of the habitués of that soup kitchen had dreams of victory soon followed by revolution, and conversation was always lively.

The aftermath of the war provided fertile ground for all kinds of would-be messiahs, and that mood seemed to extend to every activity of concern to me. A moral imperative—as we were told from many sides—was total commitment to a great and well-organized cause, or perhaps to several compatible ones. Many of the noises I had already heard before the war in Warsaw hit the older me more strongly, and they were joined by fresh ones.

Communist parties following different schools of thought were constantly competing with one another in their proselytizing. A growing number of people in France and elsewhere in Europe had come to believe that to tackle the gigantic problems ahead, democracy and individualism were outdated. They had to be sacrificed to concerted collective action—a political secular religion with a charismatic leader. Empires had been destroyed by the world wars, but most of their key structures remained popular and many countries promptly reinstated them, often in a smaller but more brutal form.

The century-old default option at Carva was to become exclusively devoted to bringing back the glory of the French state and its institutions—to become a mandarin, or *grand commis*. But after the war, many of my roommates there considered the opposite option: fearing that France would not rebuild in their lifetime, they spoke of moving to a more promising place far away—not the United States (oddly, as I thought then and continue to think today) but Argentina or Brazil. A glance at the alumni directory shows that this talk was not followed by action.

There was constant talk in 1945 of adopting some form of *sacerdoce* (vocation) or other ready-made commitment. Catholics were offered several strengthened or modernized versions of their faith. Blaming Rome for compromising too willingly, some became Calvinists. Every serious form of commitment imitated the rules that organizations like the Freemasons and the Catholic Church had themselves adopted from their predecessors. Everyone "wed a discipline"—as Jesuits wed the church and bear witness by wearing an iron ring on their finger. French society being stable, commitment was often inherited.

The celebrated writer Jean-Paul Sartre belonged to a prominent family from Alsace, the Schweitzers. A demagogue, he wrote clumsy French but had a silver tongue. I once attended a political rally that featured him. The chairman concluded by wishing to see him become a political leader. I soon came to shudder at the very thought.

The levelheaded writer Raymond Aron—whom, shamefully, I failed to appreciate until much later—was Jean-Paul Sartre's classmate at Normale. He once complained to the effect that "I have always been right but no one knows me, while Sartre is consistently wrong but famous" (well, no longer). I was urged by friends to recognize Aron as a kindred spirit, but I discounted him because he wrote for *Le Figaro,* a stolid, conservative newspaper no one I knew would touch.

The crosscurrents of that period affected me profoundly. Almost all my friends joined loud causes. The skepticism my family had instilled in me during the 1930s was amplified, and against odds—and like the Raymond Aron I missed knowing—I elected to be a dissenter. I believed that to dissent from one church, one need not create another. My ambition, my megalomania, was to help the church change. In mathematics, who was the self-appointed messiah? For better or worse, that person was André Weil of Bourbaki.

Total commitment to causes benefited some of my friends, at least for a while. But I was never tempted to join. Instead, I began to continually customize my life in a way that history might reward but society had left unfulfilled. This choice may have contributed to the length of my active life, but it also guaranteed that I would be anything but precocious. In the absence of a well-defined set of rules to play by, the very notion of precocity ceases to make sense.

One look at France after 2000 suffices to show that, in some ways, two hot wars and a cold one had little effect on the country-. Political Marxism and Gaullism—as well as their intellectual counterparts in Bourbakism, existentialism, and Freudianism—appear to have burned out . . . but who knows.

9

A (Then Rare) Foreign Student at the École Polytechnique, 1945–47

"*POUR LA PATRIE, LES SCIENCES ET LA GLOIRE*" has been Carva's motto for years. The link to the fatherland resides in the school's focus on providing the French state and society with technical elites—both civilian and military.

Carva graduation rank is perceived as extremely important and as justifying gigantic investments by both society and individuals. A very high graduation rank provides a splendid state job, which is often followed by a splendid career in business. Those aiming for the top had to work very hard, and—like in the *taupe*—efficiently, as their lives were relentlessly packed. Most of my Carva classmates found the competition excessive. They preferred to coast along, confident that being *ancien élève* (*antique* or alumnus) would serve them well in any career they chose.

The many perks that come with the degree give few graduates an incentive to live outside of France—ordinarily, a prerequisite to renown abroad. A special case might be me—of Lithuanian extraction and Polish birth. For me, at first Carva meant a great deal; I became a French citizen while a student there. But then it became less and less important until it faded into a nice memory of youth. When I graduated, I *really* finished Carva, while for the typical alum, being *ancien élève* was a welcome life sentence.

Student Life at Carva, the Military Academy

Carva was founded in 1794 as a school of civil engineering, but Napoléon made it a military academy for artillery officers and mili-

99

tary engineers, together with a few high civil servants. Some alumni were also part-time scientists who contributed mightily to the glory days of French science—1800 to 1850, later extended by Henri Poincaré (class of 1873). Science suffered a long and painful low between my teacher Paul Lévy's class (1904) and—roughly—mine (1947).

In my years, very few graduates became officers, yet the school was run like a strict military academy. Entering students immediately became state employees, and therefore had to have been French citizens for at least five years. I was a special foreign student. As mentioned, I had taken the regular exam, and I was promised a diploma if my record was better than that of the worst regular student. With the exception of a classmate who died, I was the school's only foreign student over a period of nearly ten years.

Most all the students lived in barracks. We 1945 freshmen did so on the comparatively elegant campus in town—on 5, rue Descartes, a few steps behind that majestic gate that Father had pointed out to me just after I came to Paris. As 1946 seniors, we lived in a cookie-cutter barracks called Lourcine, a good walk south of the Latin Quarter.

We were organized in units of twelve called *caserts* (short for *casernements*): tightly packed beds in a small dorm and desks in a common study room. Three *caserts* formed a "group" for recitations, gym, and foreign language classes. Serendipitously, two birds were killed with one stone—the school favored proficiency for language and a blend of talent for gym and the key subjects. Therefore, first we were ranked by our entrance grades in English or German, then we were divided into *caserts*.

Given the strict rationing on the "outside," this military academy fed us surprisingly well. Though gym was considered important, the crowded neighborhood left us limited facilities within the school. A swimming pool in a basement was so busy that we were able to use it only very early or late in the day. I still recall with dread those long jogs along the quais of the Seine under a fine drizzle before sunrise. (The joke is that Paris has no rainy season, because it rains a little bit every day.) Overcrowding became a reason for moving the school lock, stock, and barrel to a windy and remote suburb.

Carva Dress Code: Always in Uniform

Many of my otherwise conventional classmates continually grumbled about the uniform, but being the impoverished oddball that I was, I rarely complained. I had entered the school literally in rags. Léon and I piled our clothes together, and I put on my worst shoes, pants, and top. A few hours later—bliss—I tossed them all into the garbage can. This was long before dressing down created its own universal uniform. Without that dress code, the differences between rich and poor students would have been intolerably conspicuous.

The school's basic uniforms included a soldier's battledress (an accepted French word) for every day and an officer's town uniform

101

(mine lacked certain insignia). Both were khaki colored, hence the school's military academy status—though only skin-deep—was highly visible and brought some incidental perks.

In addition, very special occasions demanded *le grand U,* the grand uniform, custom-tailored in heavy black wool with long rows of gilded buttons and red and gold trim. It could be worn with either a very long coat or a billowing cape. The two-cornered hat was vaguely Napoleonic, except that the corners pointed to the front and back. Each of us also received a straight sword, which—unlike the uniforms—had to be returned upon graduation. Mine was a hundred years old, and I took it for granted that it had never touched blood.

A recent move revealed that two of the four pieces of my grand U have survived all my previous moves. I was reminded of my serial number (1179) and of the pride I had felt when—shortly after I had trashed my rags—I was measured for a masterpiece of custom tailoring. Somehow, it no longer fits—but one day it just might again.

The grand U was essential for the many elegant parties to which I was never invited and for the rather frequent parades down the Champs-Élysées. Out of a class of some two hundred, the tallest students paraded in twelve rows and twelve columns, and 144 grand Us—including mine—figured in every pre-TV newsreel. In terms of precision drills, the hardest were the few minutes when each row had to remain straight as we fanned around the Arc de Triomphe. Fortunately, the big stand with officials and guests was safely farther down, on the sunny side where the Champs-Élysées widens into a park.

To march in the first row during a parade is extremely stressful. I often failed to avoid this fate. My classmate André Giraud (1925–97), a bit shorter in stature, invariably arranged to march just behind me, admonishing me each time my drawn sword—*tangente*—deviated even a bit from the required vertical direction. He became an admired but ruthless high civil servant and, in due time, one of the few ministers (of industry and then of defense) who continue to be remembered. In school, we were anything but kindred souls, and I am glad my fate never came into his hands.

But on a much later occasion, when our paths crossed in New York, we shared a connivance that must link old dogs who recall playing together as puppies. Marching in grand U, we had performed for Charles de Gaulle, Winston Churchill, and other lesser historical figures. Most oddly—and memorably—we honored Ho Chi Minh! This Vietnamese leader was visiting Paris to iron out some remaining details of a peaceful withdrawal of French troops. That very night— without authority from Paris—Admiral Georges Thierry d'Argenlieu, the French naval commander in Indochina, bombarded the northern port of Haiphong. Ho flew back home in a rage, and the rest is history. Who was this d'Argenlieu? A naval officer who became a Carmelite monk, but received wartime leave from the order to join de Gaulle in London and was made admiral. Having defied the Vietnam policy of his country's postwar government, he returned to the Carmelites, and so could not be indicted, and was never heard from again.

Côte d'Amour and Hazing

We all received a grade for military bearing. Most officers did not want this grade to affect ranking; therefore, most students received a grade of 15/20. But there were exceptions. For example, André Giraud's first-year grade was something like 18/20—hence the grade was rudely referred to as the *côte d'amour.* My first-year grade was a lowly 2/20. The second year, I improved to a 13/20. Years later, someone having fun let me read my file. I saw that after the second year, Captain Wolf commented that although my 2/20 suggested a willful troublemaker, it only meant that I had no concept of the role of military authority. This was indeed the case—and my whole life's orbit was to show that professional authority did not awe me either.

The man who gave me the grade of 2/20 had an ax to grind. During the fall, while he was a lieutenant hanging around without obvious assignment, we (at least I) did not know that he was being groomed to take over our company. Then, over the Christmas break, he took a few of us to Fort de Briançon, near the Italian border in the Alps. Without an instructor, we were struggling to learn to ski by gliding down a highway. I almost crashed into him, making him scream, which was quite an embarrassment. Upon returning to Paris, he was promoted to my captain, and remembered everything vividly.

Gaudeamus igitur, juvenes dum sumus (While we are young, let us rejoice). In that spirit, and knowing we were going to be forced into dull and ordered lives, Carva allowed a great deal of (frankly sophomoric) tomfoolery. I barely participated, and it is well documented, so let me be brief.

Hazing of freshmen by seniors was permitted, but this was not a big deal. On a certain festive occasion, select incoming freshmen were given nicknames, sometimes mildly degrading ones. A classmate's surname, Godet, denoted a cup in French, so he became Fanofbooze and was ordered to spend an induction ceremony pretending to drink from an oversize empty cup. My name was recognized as being German for cake, something edible, so I became Fanamagnan—that is, Fanofchow—and was ordered to pretend to chew on a large cooked bone.

The Gamma Point is a special day for astronomers. Carva celebrated it with an open house. It is widely claimed that its grossly inflated prices sustained the Carva counterpart of U.S. high school student councils. Partaking of the bubbly that I was selling, I verified by repeated experiment that with good champagne it is very hard to get drunk.

Matchmaking and the Carva Alumni House

Coeducational U.S. universities are essential to matching life partners. In my time, all Carva students were men, hence the need for special arrangements such as the dances with live orchestras that the GPX— Groupe Parisien des X—hosted at Carva's alumni house in the elegant

Faubourg Saint-Germain in the Seventh Arrondissement. It had servants' quarters in a building on the road, a paved courtyard, a garden behind, and a main house *entre cour et jardin*.

Students paid no admission, and alumni acted as discreet chaperones of their eligible daughters. Many of my classmates found a wife and a father-in-law eager to become their patron. Only once, several years after graduation, did curiosity make me visit. Yet my wife, Aliette, and I held our wedding reception there, and when needed, we take advantage of its intended role as a grand home away from home.

Home Economics

Financially, the regular students at Carva (and also Normale) could fairly be described as extremely privileged—or, less politely, as utterly spoiled. This helped explain the school's attractiveness. Although it is true that those students were constrained by a long-term legal commitment to serve in the nation's army or some branch of civil service—one entirely determined by their school record—an alumnus could in fact buy back his freedom, either by paying his tuition retroactively or by performing acceptable good deeds. While at school, all regular students—rich or poor—were exempt from paying room, board, and tuition.

As a foreign student, I received a bill for tuition and board. But this was an accounting fiction: the grant from a government agency that would have gone to Normale went instead to Carva. When I became a French citizen, I lost that support, and the invoice I received was equal to the buyback of a regular student's contract. When set up in 1943, the amount was more or less equivalent to tuition and board at Yale. But by 1946, postwar inflation had reduced it to practically nothing. So, instead of making me seek another scholarship with different strings attached, Father bought me the joys and sorrows of independence—at a historically low bargain rate.

But this was not all! Every regular Carva student received a working civil servant's starting pay as pocket money. This helps answer the question I am often asked by U.S. parents or teachers: "How come twenty-year-old students in France are so much better in math?" Part of the answer: "Because they are, in effect, bribed."

This pocket money was denied to me. Bless their hearts, the elected class representatives (one of them had met me in Lyon) intervened. They were called *caissiers* because they were trusted to manage a cash box to which all the students contributed—and to which were added various windfalls. They thought that, to preserve collegiality, I should receive a comparable benefit from the student council budget.

This suggestion was voted on and defeated. Some classmates explained their nay as a matter of high principle ("You signed no obligation and therefore are owed no compensation"); other excuses sounded more like low politics ("I like you, but my friend so-and-so has objections, and I will vote as he did").

Bless their hearts—again—the *caissiers* had a higher idea of school solidarity. Their responsibilities included contributing to the neighborhood charities. The rue Mouffetard, next door, was not yet the spruced-up baby Disneyland of today, but an ancient slum. In effect, as few classmates knew, I was handled as a neighborhood charity case and granted that "benefit" anyhow.

How Did My Carva Classmates Fare?

Did all that competition pay off? Not really. Graduation rank actually predicted future performance very poorly. Yet many of my classmates played key roles in rebuilding France after the war. They faced weak competition because our immediate elders had led largely disrupted lives, were not fluent in English, and suffered other handicaps.

Low exit rank guaranteed comfort but not always a grand life—with two notable exceptions. Jean-Claude Simon (1923–2000), a roommate, gave no thought to class ranking—except that he tried to be ranked last, while managing not to flunk out. Having inherited a banking job that he found unbearably boring, he was rich enough to purchase freedom. Almost from scratch, he started a second career in

electronics and did well—first in research and then in senior management. He then had a third career as a university professor of computer science. After retiring, he built an imaginative and successful start-up that managed automated signature recognition on checks under a certain amount within installment plans. He was fun, an interesting man, and a good friend.

Another classmate, Valéry Giscard d'Estaing, stood out in school by wearing a blue uniform different from our khakis—and later by being elected president of France. I first saw him when he entered the twelve-man *casert* to which I was assigned in 1945–46. "I am looking for Simon. Do you know where he is?" "Not the foggiest." "Tell him I stopped by." When Simon came back, I did tell him, and asked who was this remarkably self-assured man wearing a different uniform. "Oh, you haven't yet met your classmate Valéry Giscard d'Estaing?" He explained the uniform and continued, "I've known him since high school. He kept telling everyone that he will be a *député* [national representative] by thirty, minister of finance by forty, president of the republic by fifty, and president of Europe by sixty. How stupid can you get?" Everyone present laughed in unison. Of course, my own ambitions may have been even wilder, but involved no schedule—and were not made public.

Amazingly, Giscard's first three youthful goals were indeed achieved—ahead of schedule. The final goal, becoming president of Europe before sixty, was missed. He remained in the public eye, as author of a European constitution. Put to the vote in France and the Netherlands, it lost. Will his dream ever be reached?

When Giscard was president of France, Jean-Claude Simon had to hand him a report he had edited. The French second-person pronoun has a familiar form, *tu,* which Simon planned to use. It is the unbreakable rule between Carva classmates and alumni from classes less than seven years apart, and he had known Giscard for years. But at the fateful moment, his mouth disobeyed his brain and uttered, *"Monsieur le Président, vous . . . "* He was crestfallen, and remained so each time he retold the story. I did not see Giscard closely again until our school's 1994 bicentennial in New York. He gave a masterful speech and we chatted, but I minded Simon's experience and kept away from the minefields.

Professors Leprince-Ringuet and Platrier

Students did not attend Carva for quality teaching, but rather for useful classmates and good jobs. Being a foreign student who didn't have to cram suited me very well and, if anything, increased my wish to excel. As a result, I received a very fine education in a broad mathematical sciences program, one that straddled the U.S. bachelor's level of the day—definitely above what I would need at the next stage of life, as a graduate student at Caltech.

My professor of physics, Louis Leprince-Ringuet (1901–2000), was a man of great charm, ambition, and energy. Fully committed to reviving experimental physics in France after its many years at a standstill, he was investigating high energies using the best tool of the day—cosmic rays. The observations were made at the Pic du Midi Observatory, in the Pyrénées near the Spanish border, and analyzed in Paris. Very popular—nicknamed Le Petit Prince after the best seller by Antoine de Saint-Exupéry—he was actively recruiting for his lab. I rushed to join his team as a part-time apprentice.

From my inherited love of gadgets and my training as a toolmaker in wartime Périgueux, I could visualize instantly—in three space dimensions plus time—the complicated instruments that the team was designing. But the rhythm of experimentation was too slow for me, and while my Keplerian plans had not yet coalesced, I was definitely bound to become a theorist of some kind.

The lecture notes of Leprince-Ringuet were uneven. On topics close to his heart, they were up-to-date, but hastily edited. Otherwise, he kept close to the notes of a Carva predecessor who had borrowed right and left. The ways of fate being inscrutable, the mysterious Carva notes made me pay special attention to thermodynamics. Even so, I didn't get it. So when I went on to Caltech in 1947, this was a course I would not miss (and thermodynamics has inspired much of my research). The Carva course had been just good enough to mystify me—and just bad enough to leave me hungry.

The Chair of Mechanics had once been held by a classmate of Jacques Hadamard, Paul Painlevé (1863–1933). After he lost the creative touch, he went into national politics—serving briefly as prime minister of France in the middle of World War I! Since the incomparable Lazare Carnot (1753–1823), I can't think of a better example of a scholar-warrior. Incidentally, his son Sadi Carnot (1796–1832) founded thermodynamics.

Painlevé continued to teach whenever he could. When he could not, his stand-in was the little-known Charles Platrier. The course and course notes changed slowly from Painlevé to Platrier, and in many small steps. Painlevé was Wilbur Wright's first passenger after Orville Wright's accident—qualifying him as a very early airplane enthusiast. The course notes Platrier prepared for my class were supplemented by many additional readings. One of them was hilarious. It contained Painlevé's pre-Wright proof that—granted certain "natural" mathematical assumptions—airplanes could not possibly fly! This proof deserves to be republished as a warning to scientists that a theory can be killed by an assumption that looks mathematically "natural" but was not chosen by nature.

Professors Julia and Lévy

Our pure mathematics teachers Gaston Julia and Paul Lévy differed in innumerable ways. When I was their student, the Paris mathematical world respected neither, and these two men and Szolem had no love for one another. This did not matter to me, and they all influenced me profoundly.

The terms "Julia set" and "Lévy process" drew blank stares when I introduced them. Today, fractalists use them every day. I was also first to use the Lévy stable processes in science, and named them Lévy flights. Although some cynics attribute to Julia or Lévy ideas that I originated, I am delighted that this terminology has taken root.

Those who closely relate to their teachers are expected to fall into a rut, and when the teachers are not fashionable, that rut is bound to be a dead end. But Julia and Lévy differed too much from each other to lead me into a single rut. Plus, all generally valid rules suffer from deviant exceptions, and I went on to prove that a person profoundly rooted in classics may very well be a successful, yet troublemaking, maverick.

Each fall Julia taught differential geometry at Carva, and each spring he was a senior professor at the Sorbonne. One course was intermediate, and the other was advanced. Double-dipping was legal, convenient, and widely practiced. A by-product is that the faculties of different institutions were not as separate as in the United States.

In 1917, Julia published his 199-page *Mémoire sur l'itération des fonctions rationnelles*. This masterpiece received the Grand Prix from the Académie des Sciences. Its topic—iteration of rational functions led to a parallel investigation by Pierre Fatou and was fashionable for a brief time. But it was filled with special examples and narrowly valid results. Bourbaki thought it was too concrete, and it fell into thirty years of scorn and neglect.

To his credit, Szolem always praised the Julia-Fatou theory, and suggested I pick it up as a Ph.D. topic. I failed to move it an inch. Who could have imagined that, thirty years later, I would revive that field with new questions that fired it with enthusiasm and brought it well-deserved glory.

Nearing sixty, Lévy was still viewed as a brilliant oddball of the first magnitude, but was "molting" into a great man in probability theory, arguably the greatest probabilist of all time. But Lévy's way of doing probability theory was too intuitive for some and too strange for others. As a result he was a loner, never to be an insider. His self-directed boldness and insight cost him much in his career and early recognition, but I found his independence admirable. I felt ready to pay the same price.

10

Pasadena: Student at Caltech
During a Golden Age, 1947–49

IN 1947, THIS WOULD-BE Kepler of complexity had reached another fork in the road. As I had hoped, Carva had granted me two years to think, and the future promised considerable freedom of choice. I learned a great deal, matured, and became very French. But freedom of choice was a negative asset; it set me on a wide sea without sufficient guidance.

I wanted to stay far away from organized physics and mathematics, and to find different, fun ways to apply my growing knowledge and gift for shape. I wanted to feel the excitement of being the first to find a degree of order in some real, concrete, and complex area where everyone else saw a lawless mess. Of bringing to a field the element of rational mathematical structure that Kepler had brought to physics several centuries before. But that Keplerian dream remained stuck in a holding pattern. I was aware that the next step after Carva was going to be hard.

Admiral Brard Recommends Caltech

In the real world of Paris and Carva in 1947, the most obvious person to ask for advice was neither Szolem nor Paul Lévy, but rather the professor of applied mathematics, Roger Brard (1907–77). A naval engineer, he held the rank of admiral and headed a large *bassin des carènes*—the lovely old-fashioned term for water tunnel. He had no office at Carva, so we met in his car. I still recall the make: Matford. A sign of the times, there were so few cars in town that he always found a parking space near the school.

In the 1930s, when the lovely SS *Normandie*, touted by *Popular Mechanics* as the latest "giant of the sea," took a trial cruise, a resonance was revealed between the hull and the propellers; Brard helped with the diagnosis and the cure. Although his numerous papers in probability theory are no longer quoted, Carva viewed him as very practical (contrary to Paul Lévy) and put him in charge of all topics in applied mathematics.

Ambitious Carva students focused solely on their graduation rank had no need for advisers. But I had a desperate need for someone with broad down-to-earth experience to help me carve a path. Brard was friendly and, to my surprise, made himself available.

With little hesitation, he made two suggestions. First, the right field for me was fluid mechanics. Second, I should go to Caltech—in Pasadena, a suburb of Los Angeles, and study under the illustrious Theodore von Kármán. Kármán was a magician who knew precisely how to find the proper mathematics to deal with great complexity. Kármán worked in aeronautics, but Brard thought that he would be open-minded.

Szolem warned me against Brard's advice. To do well teaching fluid mechanics in Paris, it was absolutely necessary to find an appropriate and reliable local patron, establish proper credentials, and only then go to Caltech. But I was restless, and none of the possible patrons in Paris claimed the magician's skills that Brard credited to Kármán.

Father viewed Caltech as an excellent idea. He had already encouraged Léon to go into aeronautics. Only later did his enthusiasm cool when he saw how close the aircraft industry was to the state.

Truth is, Father and I agreed on a plan of action but for very different reasons. I viewed aeronautics not as my final field of work but as the best available path toward reaching my Keplerian dream. So I applied to Caltech with a letter of recommendation from my Carva physics professor, Louis Leprince-Ringuet. I was accepted and spent two years there. For my travel, I got a generous stipend from Carva—arranged by Professor Brard, who had gone far beyond giving advice.

I wondered whether Father remembered that he had wanted to send Szolem to Berlin to study engineering. In both cases, engineering involved the technology that had won the previous war: from the 1920s to the 1940s, it had changed from chemical to aeronautical.

Father could not possibly have heard of the advice received from the fathers of three famous Hungarians: the mathematician John von Neumann, who will play a large role in this story, and his contemporaries, the physicists Edward Teller and Eugene Wigner. Their fathers—far more prosperous and worldly than mine—had also insisted that their sons study chemical engineering. So they did, with profound historical consequences when they worked for the U.S. government during World War II.

Welcome to Los Angeles

In terms of direct preparation for a career, I pretty much wasted my two years at Caltech, though some courses came in handy later. However, my time there gave me the chance to refine my Keplerian dream. I am very lucky to have gone.

The only crossing I could book was from Southampton to New York on the SS *Queen Elizabeth*. It had recently been converted from a troop transport back to a luxury liner—though my tiny shared cabin on the lowest deck was grim.

I managed a sightseeing stop in London and reached New York dangerously close to the beginning of classes at Caltech. Someone paid for an air ticket, and I flew to Los Angeles. The limousine from Manhattan to the airport stopped next to an opening in a big wire fence. Right on the other side stood a gleaming silver plane, an early model Lockheed Constellation of Trans World Airlines. Its four propeller-driven engines could not reach Los Angeles without a stop halfway at the TWA hub in St. Louis. Tickets were checked by one of several employees idling at the gate, and off we went. This was my first acquaintance with what is now New York's La Guardia Airport.

My first impressions of California involved smog and the Bible. For days after I arrived, my eyes hurt uncontrollably. I recalled that a man to whom I had been recommended by friends was an ophthalmologist. He lived west of downtown, much too far for social interaction, but I called him for professional help. He gave me an appointment at no cost, but on the way I missed the big red streetcar along the Arroyo Seco. Desperate, I hitchhiked. I was soon picked up by a two-door sedan with a young driver and an older passenger. I sat in the back. Once the car started, the passenger turned to look at me and inquired, "Are you saved?" Wondering whether my ears were also affected by Los Angeles, I failed to respond. The car swerved left and right, and the passenger turned toward me again. "My son is a safe driver, but accidents often happen on the Arroyo. Think again, are you saved?" At that point, the car stopped on the side of the road, and the passenger moved to the backseat next to me. He opened his Bible and read a passage. "Does not this story read the same in English as in French? Don't you agree that this proves the existence of God? Think again, are you saved?" At that point, I got out and they drove down a side road, leaving me stranded. A streetcar finally arrived, but I was very late for my appointment. A message taped on the door informed me that the doctor could not wait. My response informed him that I was very sorry but had been delayed by a preacher.

When I finally saw the ophthalmologist, he greeted me and asked, "Was that preacher any good?" His medical diagnosis was that my eyes were fine except for being overly sensitive to the smog. "What is smog?" "Oh, you had not been told? It is smoke mixed with fog. It's part of the weather here. Some of your friends at Caltech are working on it. Ask them." I did.

Another very different surprise met me at the first room I rented. The landlady and her friends spoke German to each other! Their ancestors left for America after Prussia defeated its liberals in 1848.

American Academia in Rapid Transition

When I checked in, the admissions office at Caltech told me, "The yearly tuition is six hundred dollars. We have not mentioned it because you won't have to pay anything. It has been taken care of by a benefactor of the institute who is interested in international cooperation. He lives close by, in San Marino. Perhaps you should send him a note of thanks."

Mea maxima culpa: I didn't. Worse, I forgot the benefactor's name.

As you may recall, not only were tuition, room, and board free at Carva, but students actually received a stipend, even those who bragged of millionaire parents. The very existence of tuition at Caltech seemed abnormal to me. My feeling of guilt did not dissipate until my sons went to college. I could afford the tuition, and I paid it instead of pressing them to seek scholarships. So my debt to Caltech was repaid to Yale and Harvard.

Far more important than tuition, the Caltech course catalog and faculty directory, hot off the press, overwhelmed me with disappointment. All too many of the stars who had made the older catalog so attractive were gone. The specifics varied from case to case. Hugely disappointing, was that the physicist J. Robert Oppenheimer—of wartime Los Alamos fame—had moved to the Institute for Advanced Study in Princeton, New Jersey, where we would meet in 1953. More generally, by Caltech's past and future standards, theoretical physics was at a low point.

The Caltech faculty was undergoing a complete overhaul, accompanied by a shift of emphasis. The reason was that the school's de facto founder and for many years de facto president, Robert A. Millikan (1868–1953), had built Caltech by bringing in many friends in the same age bracket. They were now all retiring or otherwise gone. Freshly retired himself, Millikan was bored and available to students for lunch and chats. These took place at a club called the Athenaeum; one of the big assets of Caltech in that both faculty and graduate students are welcome. Flashing ahead, I was once having lunch there with Millikan when a wan and shabbily dressed gentleman came forward and, very formally, bowed and introduced himself as Laue. I recognized a true giant of physics, Nobel in 1912, whom Wilhelmine Germany made Max *von* Laue. He had not bent to Hitler, but sad to say, Millikan treated him imperiously nonetheless. I felt and continue to feel that good Germans deserved better.

The math offerings at Caltech were limited. The one professor with any name recognition was Eric Temple Bell, though it was largely earned for a collection of biographical vignettes of individual masters titled *Men of Mathematics*. That book was both blamed for historical howlers and credited with enriching the field with many enthusiasts. Intrigued by the man, I often attended his Sunday afternoon open houses just across the street from Caltech in a neighborhood the faculty liked and could afford. A crusty Briton, he kept attacking the much-discussed proposals for federal funding of scientific research through the National Science Foundation. He viewed all federal support as a potential threat to existing local collegiality and the proper decentralization of the U.S. decision-making system—a move toward the horrors of French-style centrally controlled departments but scattered locations. He was farsighted, but I disagreed until much later, when the NSF had consolidated into a bureaucracy from which a solo throwback like me could receive only crumbs.

Fluid Dynamics in a Period of Maturity

The worst disappointment of all was that, although the Caltech catalog continued to list Theodore von Kármán as active faculty, he was on leave and had taken up residence in Paris! He had never married, and his sister—also unmarried—managed his household. They had lived in Belgium while he was a professor at Aachen, just across the border in Germany, and she accompanied him to Caltech. But as soon as the war was over, she wanted to return to Europe, where she settled in an elegant hotel in Paris. He dropped by Caltech several times during my stay, then retired.

Even worse was the replacement of the famed "Kármán circus" by people hardly anyone had (yet) heard of. Fluid mechanics as a whole had become an extremely competitive and "mature" field that was growing slowly and splitting. The rules of fluid motion—called Navier-Stokes equations—are of infamous difficulty. Pure mathematicians and physicists had little to contribute, so it was left to engineers. Another problem: one of their leading textbooks was written nearly a hundred years earlier by an English don, Horace Lamb, before he became Sir Horace.

In 1947, a key research question was what would happen if an airplane could accelerate enough to reach the sound barrier. Ivory tower theoreticians agonized in one world, and adventurers in another made immense amounts of money to fly unproven rocket-powered contraptions that might or might not take off, fly, or land safely. Kár-mán's vaunted combination of theory and practice was no longer a glue that held the worlds together.

One saving grace was that while the dynamics of smooth flows had matured, the study of turbulence had not. In fact, it was only just beginning to reveal its devilish complication. Story has it that when the great physicist Enrico Fermi (1901–54) was close to death, his friends wanted to know what his first question would be when he met his maker. "What is the cause and nature of turbulence?" was Fermi's response. In other words, "What is the essence of Navier-Stokes?" Altogether, watching Caltech colleagues at work on turbulence and so-called spectral, or harmonic, analysis was to serve me well in the future.

Living beings mature, then age and die. However, a science may very well move back from sagging maturity to wild youth. This is what happened to general relativity theory, a mature field in 1950, and to the mathematics beloved by Szolem. Later, the theory of chaos contributed to fluid dynamics and brought me back to it for an important effort: developing a concept called multifractals.

Tolman, Liepmann, and Other Unforgettable Courses

Not surprisingly, the Caltech curriculum was a letdown after Carva. A course in elasticity was compulsory, but the one I had at Carva was more advanced; therefore, I dared to cut many lectures. My final exam received an E—instead of solving the problem, I showed that it could not be solved because something was missing. I protested and had (in effect) to teach some delicate points to the lecturer. He relented—but refused me an A.

The course work reached a high point with the swan song lectures of Richard Chase Tolman (1881–1948) on statistical physics, or thermodynamics, an extraordinarily difficult and subtle topic that brings many seasoned scientists to beg for help, run away, or make dreadful mistakes. Tolman was no technical acrobat and was about to retire, but he started the class with a warning—his course was for those who already knew the subject and was not intended to teach thermodynamics. But he promised to show why it worked. That he did and explained away many of the mysteries that had baffled me at Carva. Learning about this topic from a seasoned master affected my work over much of my life, and helped my thesis and some papers add a few wrinkles to the logical foundations of thermodynamics.

After Tolman, I learned most from the fluid mechanics course of Hans Wolfgang Liepmann (1914–2009). He *did not* put the stress on formalism, but on proper understanding of the physics. Once, when he was criticized for his harshness, I heard him say that "simply being Jewish does not prevent me from being a true-blooded Prussian." He was the only professor I ever feared.

I also recall fondly two teachers outside of science. The pride Caltech derived from requiring courses in the humanities was well justified. Wallace Sterling, who taught Shakespeare, had been moonlighting from the Huntington Library next door and was also a respected radio columnist. He soon left to lead Stanford University to its present high distinction. Horace Gilbert taught a course on economic institutions. His staunchly conservative stance was not to my taste, but the class was fun and I learned a great deal.

Airplane Design with Paul MacCready

On the practical side, I have a fond recollection of a part-time instructor named Klein. A Caltech physicist before the war, he had become a personal technical consultant of sorts for Donald Douglas, the founder and namesake of a then-flourishing aircraft company. This day job made him arrive late and totally unprepared, and he would entertain us with gory stories from the real world. For example, the ads boasted that automobiles were becoming a bit wider every year. Klein told us why: the stamping machines of that day wore out fast. The cheapest solution was to replace the hard surface of the "male" part and gouge the softer "female" side. Klein's stories kept alive for me Father's love of machines and gadgets.

A far more important requirement challenged us: to design a refueling jet tanker—minus the engines. We split into teams of four to deal respectively with the fuselage, wings, tail, and landing gear—

which I was in charge of. World War II planes—like the workhorse DC-3—had a small fixed wheel near the tail. It was easy to design, but on the ground made the plane lean back awkwardly. Big retractable wheels in the front—like on the DC-4—made the planes horizontal on the ground, but still posed many design challenges.

Our specific design was not memorable. Of course, the aircraft companies' experts were reading the same books and magazines we were. So when the tanker precursor of the Boeing 707 came out, I recognized "our" approach—and still follow the evolution of its successors. Passenger airplanes have greatly improved in every detail—but not overall. This gives continuing evidence that airplane design has been since 1947 a "mature" field. Spending a lifetime on such details would have been a dreadful experience, and did not tempt me at any point.

One teammate, Paul MacCready (1925–2007), was unforgettable, and despite our many differences, we were friends. He became an imaginative, tireless, and thoroughly old-fashioned independent inventor whose company, AeroVironment, was motivated by curiosity rather than profit. His fame soared when he designed "airplanes" in which a (well-trained) pilot's leg power was transferred to a propeller through the pedals and gearbox of a racing bicycle. MacCready's fascination with bird flight was evidenced by the names he selected for his aircraft. The Gossamer Condor tested the concept, then the Gossamer Albatross crossed the English Channel at low altitude—just above a flotilla ready for any emergency. The later Gossamer Penguin used solar power, as did the Solar Challenger, which crossed the channel at a higher altitude.

But all that fun came later. At Caltech, he spent his weekends practicing soaring—that is, imitating the birds by flying gliders, light planes with no engine. Only later, when I was serving in the French air force, did I hear from pilots that this soft-spoken man had been the U.S. soaring champion many years in a row before becoming the world champion. A wider community named him engineer of the century. Calling himself an ambivalent Luddite, he advocated unbridled thought. He was a lucky bastard who managed to never leave the sandlot and just kept playing. We always felt in tune but did not see each other often enough. A splendid man.

The Mathematical Faces of Mechanics

At Caltech, one of the mathematical faces of mechanics was represented by my glorified master's thesis. Frank E. Marble assigned a topic in propeller theory, and I worked out the complicated calculations—without either of us becoming too involved. Frank and I remained friends, and he boasts that he helped save me for higher pursuits.

Another face was personified by the mathematician Paco Axel Lagerstrom (1915–89). A brilliant and cultured Swede, he was strange and mysterious—admired by some students, tolerated by a few, resented by many. We often met socially, so I learned that Paco's evolving taste had moved him from divinity to philosophy, logic, very pure mathematics, and then a flavor of mathematics he viewed as applied—but I didn't. On a rare visit, Kármán asked me to describe the topics I was considering for my dissertation. Soon after I began, he broke in to demand which fool could have suggested such an antiphysical topic. I had no choice but to point at Paco, standing beside me. He was called on in turn, and Kármán did not treat him kindly.

After that incident, my relationship with Paco deteriorated. A course I was taking with him ended with an oral exam. He gave me an A, and then told me, "I think you should not start on a Ph.D. with me because you don't admire me enough." He was right and I appreciated his bluntness. Inertia might have led me to try working with him, but the sequel would have been either brief or regrettable. Unfortunately, he was the only professor at Caltech I could ask to supervise my doctoral work. This meant leaving Caltech without a doctorate.

In a defeated mood, I for a moment yielded to the force of gravity that continued to pull me to pure mathematics. I rejoiced when a top math department, at the University of Chicago, seemed to offer an assistantship. But that assistantship was actually not funded. They said

I should register anyhow, because the great mathematician Saunders Mac Lane badly needed a teaching assistant for his algebra course and would find a way to support me. Algebra was (and remains) my least favorite topic in mathematics, and I was not ready to face uncertainty for its sake.

No Doctorate, but Quality Education and Community

First and foremost, it is through Caltech that I met Aliette—although this meeting and our marriage did not occur until years later.

More broadly, the small crowd that chance or necessity had collected at Caltech during my time there was of a quality that I rarely saw anywhere. The intellectual excitement and the feeling of living through uniquely exceptional times were palpable and exhilarating, though burdensome, and persist in my mind to this day. This is also how the world came to feel, since that very small school, over those

very few years, produced astounding numbers of Nobel Prize winners and the like. Also, Caltech may be one of a kind in not increasing its permanent faculty.

I had great fun, enjoyed the Southern California outdoors, and made many lifelong friends. The physicist Donald Glaser and I often went to concerts together, and I followed his career closely. As an experimental high-energy physicist, he invented the bubble chamber, which experts in thermodynamics had declared impossible—against the laws of physics because it contradicted a statement in a book by established physicist Enrico Fermi. That statement was revealed to be incorrect, the chamber became a basic tool, and Glaser earned a Nobel for it. Only then did he reveal that he had "converted" from physics to molecular biology. True to form, he attended the meeting on my seventieth birthday celebrating *my* versatility and regaled the participants with tales of *his*.

Particularly important was Caltech's Inter-Nations Association, which attracted foreign students and young people from town. The school assisted the INA—possibly even sponsored it. We learned about the New World and described the old one to young Americans who had not yet seen it for themselves.

One INA regular was Paolo Comba, a math student from a Protestant corner of Italy. Our paths crossed again when we were both at IBM. In retirement he went on to discover many minor planets and marked an old friendship by discreetly naming one after me. As a young scientist, he worked on baby tomatoes, predicting they would move fast from the lab to the grocery store. They did.

Caltech confirmed my cynicism about an opinion taken for granted at Carva—that a school's elite status is largely based on attracting elite students. My age cohort included several elite students, but more important were the many with complicated, often heroic backgrounds in wartime.

Max Delbrück and the Birth of Molecular Biology

On the small Caltech campus, the burning center of intellectual life was not found in aeronautics. It was a group led by a man of ambition, brilliance, and independence of mind: the great maverick Max Delbrück (1906–81).

After an inconsequential year in mathematics and aeronautics, I met Gunther Stent (1924–2008), who was at that time a physical chemist. He introduced himself as an incoming postdoc with Delbrück and told me that in a few days another postdoc, the microbiologist Elie Wollman (1917–2008), would arrive from the Pasteur Institute in Paris, along with his wife, Odile. In no time, Gunther and the Wollmans became lifelong friends. Soon I met the phenomenal D. Carleton Gajdusek (1923–2008), an academic superstar. I had moved socially and intellectually over to the in crowd.

Then and there, despite the reservations and declared hostility of several well-established guilds, Delbrück was orchestrating the birth of a new way of being a biologist. At Caltech at that time, the word "biophysics" was forbidden. But soon their work would become "molecular biology." And in 1952 this field would come to be known universally, in response to the discovery of that icon of natural geometry—the double helix of DNA. Eventually, molecular biology merged with biochemistry, and genomics took it to an industrial stage. Today's practitioners complain of it being viewed as a mature field. But in 1949, nothing was further removed from the slow-moving maturity of fluid mechanics.

A high Prussian aristocrat—a Junker—Delbrück had to leave Germany because he would not swear allegiance to Hitler. One of the would-be assassins of the Führer was a cousin of his. How did he manage his unprecedented transition from physics to biology? His early years were unspectacular. As a physicist, he felt that he was hopelessly behind Hans Bethe (1906–2005) and Victor Weisskopf (1908–2002)—his near contemporaries in the entourage of Wolfgang Pauli (1900–58)—and a Delbrück would not settle for second best. The epoch-marking lecture "Light and Life," which Niels Bohr (1885–1962) gave in 1932, sparked him to become a biologist.

With Salvador Luria (1912–91), later a Nobel laureate with him, Delbrück wrote a paper that Erwin Schrödinger (1887–1961) noticed and mentioned conspicuously in his book, *What Is Life?* When the war ended, Caltech went recruiting and gave Delbrück his first real job as a full professor of biology. In time, crowds of physicists followed Delbrück's new path into biology and the once-spurned biophysics won acceptance. So what did he do after his field became established and ubiquitous? True to his temperament, he left this field for another far less explored one.

A Belated Delbrück "Treatment"

Delbrück's personality would never be described as mild. One day, I noticed that our parties were no longer graced by the presence of a man whom I only recall as Harold. Having asked, I was told that Harold had received the "treatment," did not do well, and was gone.

The Delbrück treatment remained a mystery until I ended up receiving it myself many years later. In 1979, the physicist Richard P. Feynman (1918–88) invited me to return to Caltech to give a lecture on fractals (this was right before I discovered the Mandelbrot set). He and Delbrück sat next to each other just under my nose. Throughout, Feynman nodded and smiled approvingly. Delbrück remained stone-faced, and as we walked out, he said casually, "Benoit, won't you come to my office tomorrow morning at eight."

On my way out, the long corridor I went through was lined with Caltech undergraduates. I stopped to ask if they had any questions. No, they just wanted to see me up close. In earlier years, I too had stood in line to catch sight of a prominent lecturer. Now that prominent lecturer was me—made familiar to those students by my 1977 book!

The next morning, I stopped by Delbrück's office. He greeted me with, "Yesterday you mentioned the name Hausdorff. Tell me more about him so I can check if he was a man I have met. . . . You said such and such. I didn't understand. Say it better. . . . Someone asked this or that. Your answer was weak. Can you do better now?" I suddenly realized that I was receiving the treatment—and was fielding each question and surviving. After the ordeal ended, he relaxed in his chair and, in a completely different tone of voice, concluded, "It was a very nice lecture. I learned a great deal."

My Keplerian Dream Acquires a Bit of Focus

To have witnessed the birth of a field from close by was an experience I never forgot. It provided exhilarating proof that someone with my bent might have a chance after all. There was much talk of physics having dominated the first half of the twentieth century, leaving the second half for biology. Even Richard Feynman tried his hand in Delbrück's lab. I never seriously thought of moving over, but I felt energized and kept looking for analogous openings closer to my strengths.

The timing was ideal because several new developments that had been "bottled up" by war conditions were being revealed in a kind of fireworks I saw on no other occasion. My restless curiosity led me to read works that were widely discussed when they appeared: *Mathematical Theory of Communication* by Claude Shannon, *Cybernetics, or Control and Communication in the Animal and the Machine* by Norbert Wiener, and *Theory of Games and Economic Behavior* by John von Neumann and Oskar Morgenstern.

Except for a fleeting thought that I might return to mathematics in 1949 via the University of Chicago, I was beginning to think that the examples of Wiener and von Neumann might guide me to an idea big enough to make me, in some way, the Delbrück of a new field. This is precisely what I set off to do.

But not immediately. I took the bus to New York, stopping to visit museums in Detroit and Cleveland. Next I took a boat and train to Paris—and fell into the open arms of the French air force, where I was to spend the next year.

11

French Air Force Engineers
Reserve Officer in Training, 1949–50

A BLESSING THROUGHOUT LIFE: I never wonder who I am. To the contrary, many successive bureaucracies wondered endlessly. The French army was certainly one of them. In trying to make sense of who I was, it improvised arrangements that had never been needed before and were likely never to be needed again.

Remember that Carva exam hell? It ended in January 1945, but the school's buildings were filled by returning veterans, so classes did not start until October of that year. My classmates took six months of basic training in a special military unit and were cleared of further military obligations. But being a Polish citizen, I was not called to serve. I tried to volunteer with my class but was told that foreign nationals could only join the dubious Foreign Legion. Bureaucrats tried to get me to join, but I was not persuaded.

When I returned to France from Caltech in 1949, Léon was graduating, ending his student deferment, and about to be drafted for a year by the Air Force Engineers. It seemed right to check where I stood with the army, and my overeagerness revealed a Gordian knot that took a year to untangle.

Carva being nominally a military academy, we students were locked in most of the time. When men in my class were called by the draft board, the authorities, knowing that they were at Carva, simply said, "This will do." As a result, my military record was stamped *"Bon pour le service"* (meaning active duty), but they did not bother to call me. Later, having discovered that I was a foreign student—and hence a civilian—they declared me a *bon absent*. *Bon* because—lacking evidence to the contrary—they deemed me fit for service. *Absent* because I had not shown up; this was the lowest level of being a deserter.

Had I not inquired, the contradiction would have remained buried in already ancient files. After I did, it was solved by deciding that, in effect, I had been granted a student deferment and should now be called to serve for twelve months.

Air Force Camps: La Folie and Château Bougon

The dictionary defines *folie* as "madness," but the term also denoted the mini palaces that eighteenth-century aristocratic ladies built to entertain their very close guests. One was located in Nanterre, a northwest suburb of Paris. By 1949, that lady's palace had become an air force camp called Camp de la Folie. Today it is the campus of the Université de Paris–Nanterre, where the famous student protests started in May 1968.

The French air force ordered me to report to Nanterre. My first requirement when I arrived was to fill out a questionnaire about my qualifications. I was prompted by a young farmer.

"Can you read and write?"

"Yes."

"Did you finish elementary school?"

"Yes."

"Did you pass the *certificat d'etudes*?"

"Yes."

"Did you go to high school?"

"Yes."

"Did you finish high school?"

"Yes."

"Anything else to report, Mr. Smart Aleck?"

"Yes."

"What?"

"I graduated from the École Polytechnique."

The fellow became red in the face and boomed, "And I am the Virgin Mary! Don't you city slickers lie to me."

"But I don't. Everything I tell you is absolutely true."

"If it were true, you would not be drafted as a private in rags, but as an officer giving orders."

In short time, I was taken to the colonel commanding the camp. "Everybody says that you claim to be *ancien* Carva. Of course, nobody believes this is true, but the sergeants don't dare order you around, and your presence creates *un bordel*. Carva alums are *never* drafted here. This must be cleared up."

"Colonel, I am *ancien* Carva. Look up the yearbook or call the school."

"OK, I believe you." He became thoughtful and said that my being drafted was obviously the result of an administrative mistake. His good friends at the air force headquarters would fix everything in no time.

When I reported again, he was subdued. "The law is the law and you must serve for a year, but certainly not as a private. You will be trained as a reserve officer and start as an *aspirant de réserve* [aspiring reserve officer]. In six months, they will call you to headquarters, but you must begin at a suitable camp—*not here!*"

The colonel's secretary broke in: "To be accepted as an *aspirant de réserve,* one must either have taken ROTC or pass a special exam." The colonel's response was, "For a Carva alum to be asked to take this exam would be completely undignified." So they added to the regulations that the ROTC requirement was automatically satisfied if one was a civilian alumnus in good standing of a French military academy.

Next the colonel was reminded by his secretary that promotion to *aspirant* required a formal appointment. Another call to headquarters, and a letter announced that the regulations had again been changed. "Now he fulfills all the requirements and shall be made an *aspirant.*"

My military record was updated and I received a new uniform, a huge backdated raise, and a one-way railroad ticket from Camp de la Folie to Camp de Château Bougon, near Nantes, to report to the captain at base headquarters.

I introduced myself to the captain. He was barely five feet tall and hated all six-footers, especially low-ranked ones. He asked to see my papers. "This letter simply *announces* that you *shall* be appointed. Only the president of France can sign those papers."

"I am sorry, but the office in Nanterre did not think the president's letter was needed. As you see, they have already updated my military record."

"This is getting out of hand. Come back tomorrow."

"*À vos ordres, capitaine.*"

The next day, the captain described a compromise. To update my record by demoting me, only to reappoint me in a few days or weeks, would be hard to explain. So I was ordered to keep out of sight and wait for the president's letter. I did, and joined Léon. I soon became an expert on the muscadet wine grown near Château Bougon—dry and dangerously cheap from the barrel.

The president's decree arrived shortly, and they sent me off for training. That training called on the good eye and steady hand that had helped me learn to be a toolmaker in 1943. I became an excellent sharpshooter, a skill I am glad never had to be tested further.

Camp de Cazaux and a Tutoring Arrangement

The day after my arrival at Cazaux, I reported to camp headquarters. The colonel there asked me, "Are you familiar with who I am?" "Yes. During the war, you were a famous fighter pilot." "Exactly. And did you notice two pilots doing acrobatics at noon?" "Yes, I noticed." "What do you think of those pilots?" "Caltech taught me that slow rolls are unstable; those pilots are nuts."

The colonel informed me that he was one of those pilots. "But don't you worry—we know what we are doing. Besides, these are World War I planes built in 1924—six years after they were needed— when they finally knew how to make them. Made of wood, cloth, and glue, but steady as rocks."

"I am pleased to hear that, Colonel."

"You studied aeronautics in America and can help me. The brass gave me all these decorations, but they refuse to make me a general because I did not go to the air academy and don't have a high enough degree. I must become a scholar of supersonic flight, but I know nothing of it. Would you agree to review my papers and tell me sincerely if you find anything dubious or wrong?"

"I shall be honored, Colonel."

He gave me his papers, and shortly after I reported back. "How is it? Tell me the whole truth."

"Colonel, this is a good beginning, but more work is needed."

Revised papers came in, and I went to see him again. "How is it now?"

"Getting there. You could include this and that."

"Marvelous. You are very helpful and will be rewarded. Those planes we fly are two-seaters. You will sit in front. First you will be sick like a dog, and then you will have a high unlike any other. You will see."

"À vos ordres, colonel."

Another revision came in. "How is it?"

"Well, actually you are backsliding a bit." I went on being picky, until my basic training was complete and I was to be packed off to headquarters in Paris. One hour before my train, I returned the last assignment to my student. "Colonel, your piece is beautiful now. When they make you general, I would be honored to attend."

"Many thanks. Come tomorrow for the reward I promised."

"Unfortunately, at noon today I shall be boarding the train to Paris."

"That's too bad. You must come back soon."

"À vos ordres, colonel." I never heard from him, or of him, again.

Paris Headquarters on the Boulevard Victor

My next assignment was at the Office of Scientific Research on the boulevard des Maréchaux, a ring road around Paris that honors Napoléon's closest helpers. The exact location was the boulevard Victor—a good name for a marshal, a good address for a headquarters, and a good last stop for a military "career" that had started in the Camp de la Folie.

My colonel had heard about me and chose me to be his scientific liaison to academia. I liaised with abandon and everyone was delighted. Incidentally, I did not wear a uniform and lived at home. I wonder what living quarters would have been available had I not been a Parisian.

A few years later, I overheard a friend enthusing about his military assignment: "One thing I can tell you guys is that the Office of Scientific Research is very civilized." Turning to me, he continued, "It seems that you don't believe me!" I responded, "Of course I do. I tailored it to my needs, and am delighted that they also fit yours."

In a serious vein, liaising was a good opportunity to scout for Ph.D. topics. At Caltech, I had read the seed papers in which Claude Shannon founded information theory, and I badly wanted to know more. A get-together in London on this topic attracted me greatly, so I asked if I could attend. The air force obliged and sent me there. It was my first scientific conference.

An Extended Sentence?

The end of my twelve months of duty was approaching, and I was counting days. But at the last moment, to show solidarity for the U.S. effort in Korea, France extended the length of compulsory military service to eighteen months. The law excluded draftees who had been deferred as students—like Léon. I thought this clause also applied to me, but the day scheduled to be my last in the air force came and went, and no one called me to let me go.

I inquired and was sent to the colonel. "Thank you for coming. The news is not good. As you know, your military record is . . . um . . . unusual. We treated it as if you had been deferred as a student, but nowhere does your record say that. We have been reviewing your case for days and are looking for a solution, but we cannot find one. The law states that you will have to serve another six months."

"But . . ."

"Very sorry!"

Fighting panic, I took matters into my own hands and rushed to Carva for assistance. The major in charge had been a captain in my time. He promptly found the carbon copy of a letter from the general commanding the École Polytechnique to the general commanding the armed forces in Paris. They knew each other, and the letter said: "Dear Friend. A graduating student, Benoit Mandelbrot, needs an exit visa to take a scholarship in the United States. His military record looks ridiculously complicated. I take it upon myself to inform the exit visa people that everything is under control and will be fixed shortly."

With a certified copy, I rushed back to headquarters. "Marvelous. That is all we need. Everybody agrees that the difficulty raised by your case was inadvertent. The rule extending service to eighteen months will be rewritten properly, and we have been authorized to let you go immediately."

In record time, I became an air force reserve lieutenant, junior grade, packed my few belongings, and walked out onto the boulevard Victor to face an altogether different set of challenges. Legal advice would probably have prevented this lost year—but I truly believe it helped me grow up.

12

Growing Addiction to Classical Music, Voice, and Opera

I HAD NO TIME FOR MUSIC until a Carva roommate, Yves Charpentier, invited me to join him at a public rehearsal. The Orchestre des Concerts du Conservatoire—today's Orchestre de Paris—was mostly made up of musicians from the Opéra who played together on Sundays in the Théâtre des Champs-Élysées. Their Saturday morning rehearsals were open to the public for a small charge. Charpentier was a regular and he liked company. And there, as he pointed out, our Carva uniforms received admiring glances.

Curiosity and loneliness made me accept—and I was hooked for life. Beethoven's symphonies—which I had not heard until I was twenty—were a revelation beyond words. At the second concert, the great Bruno Walter (1876–1962) conducted the Fifth, so my baptism was high-class indeed. Only a few weeks later, Charpentier declared that, having started as a complete neophyte, I had absorbed all I heard, as a dry and thirsty sponge absorbs water. In no time, I had become more knowledgeable and learned than he—who had been listening all his life.

Charpentier also introduced me to Carva's music room, with its sizable collection of old 78 rpm records. Hard to believe today, but the record player's needles were made of the same wood as the reeds of wind instruments. Lighter than metal, wooden needles had to be continually sharpened using razor blades; they scratched the records and did not last long.

I owe an immense debt to Charpentier. I thought we could become close friends. But one day, he vanished without a word. I guess that, except as music lovers, we had little in common; in fact, though he was a Parisian, I never met his family.

* * *

After arriving at Caltech, I discovered the public "concerts"—records played in a large lounge. Next to it, a good-size control room was nearly filled with huge boxes—the top professional hi-fi equipment of the day—and shelves groaning under the weight of 78s. Three visitors were a crowd; mostly I shared that room with John McCarthy, who was to become a founder of computer science. An outspoken left-wing political extremist, he criticized Henry Wallace for timidity yet wanted him to become president in 1948. Eventually, life led him to the extreme right. In choosing what to play, John and I had to compromise. I am grateful he forced me to listen to Mahler.

In Pasadena, I heard the pianist Vladimir Horowitz (1903–89). Shortly after one of his notorious and long "intermissions," the big hall was quite empty! Astonishing technique, but listening to him made me fidgety and restless.

Far more satisfying was a recital by the unknown Rosalyn Tureck (1914–2003). A "brainy" and profound interpreter, she played Bach on the piano, a marvelous maverick. Years later, we became friends, and I told her about this Caltech concert. She remembered it perfectly as a turning point. For the first time in her life, the crowd treated her not as an oddball but as a pioneer.

While I was not yet enamored of the human voice, I heard the great diva Lotte Lehmann on one of her "last" tours. As she performed Schubert's *An die Musik,* her voice cracked, and she stopped and apologized. Many elderly people in the audience were crying—but I confess wondering if she was not simply behaving like divas are supposed to during their farewell tour(s).

* * *

At the Office of Scientific Research, my colonel's secretary, Françoise Mer, was by no means a flawless record keeper or typist, but she was a cultured and musical upper-class lady who needed the money and liked to discuss opera with me. I had become a chamber music fan but knew almost nothing of opera. The few opera records at Caltech included the unsurpassed prewar Glyndebourne Mozart recordings conducted by Sir Thomas Beecham or Fritz Busch. Françoise conceded that my special admiration for the bass-baritone John Brownlee in the title role of Mozart's *Don Giovanni* showed good taste. Long before, I had of course loved *Carmen,* as sung by the tenor Georges Thill, master of a vanished French singing style.

Developing a fitting awareness of the marvels of the human voice became a priority. Somehow, I was assigned for a week to an air base near Aix-en-Provence during the newly founded music festival, modeled after Salzburg's, which featured opera. Next came a weeklong assignment near the newly revived Salzburg Music Festival, where I heard great performers. Yehudi Menuhin played Bach's Chaconne for Violin for a score of music students in a very small and ornate room. Wilhelm Furtwängler led the Vienna Philharmonic in a Bach Brandenburg Concerto, conducting from the harpsichord and including his own (strange) cadenza. A young woman ran down the stairs and out of the building, whistling beautifully, and I recognized the famous soprano Irmgard Seefried. Below is a picture of me in Salzburg during that most pleasant and educational week—more than I could have ever imagined from the military. Later, I stopped in Vienna, went to

the opera, and heard *Carmen,* with another great soprano, Hilde Gueden, in a minor role.

Opera became a passion. A true opera nut remembers the best performances to his dying day. I soon became a demanding expert on singing in general. One day the radio was broadcasting from Toulouse a stunning concert by the unknown soprano, Victoria de los Ángeles. The announcer mentioned that she was to sing the next day in Paris. I rushed to the Salle Gaveau, bought the cheapest ticket, and got a choice seat in the orchestra. Why? Only a dozen people attended—half of them recognizable artists. When she sang a few months later at the Théâtre des Champs-Élysées, it was jammed.

The ancient, elongated Salle du Conservatoire—a few blocks from home—never sold out. I often dropped in for concerts and "discovered" several other future greats. A young and skinny (!) flutist named Jean-Pierre Rampal played beautifully to a near-empty hall. I also heard the venerable George Enescu—the legendary teacher of Yehudi Menuhin—play in a hall so packed that I was seated on the side of the stage. Bent and arthritic, he held his instrument straight down. To accompany him might have been unfeasible, and indeed I recall no accompanist. He too played the Bach Chaconne (!!) to rapturous fans crying in unison. Of course.

In mid-twentieth-century Paris, the prevailing taste in music was not daring at all: Claude Debussy and Maurice Ravel (long dead)—not to mention Igor Stravinsky—were still widely viewed as wild modernists. This helps explain the virulence of the French musical avant-garde that was to be exemplified by my contemporary Pierre Boulez.

Since then, I have become attuned to more way-out current fare. I boast the composer Charles Wuorinen as a friend and was close to composer György Ligeti, who died in 2006. What brought the three of us together was a special development—the observation that music has a fractal aspect.

13

Life as a Grad Student and Philips Electronics Employee, 1950–52

IN 1950, I BECAME A NOT-SO-YOUNG mathematics student at the University of Paris in search of a good topic for a doctoral dissertation. Unlike today, Carva did not grant doctorates, so I went to the University of Paris—then at a low point in its long and often glorious history. Its requirements for the doctorate had not changed in years, and shortly afterward were made stricter. The course requirements were minimal, and I had fulfilled them with no sweat in 1947. French academia was about to be pushed from ancient anachronism and immobility into perpetual modernization. These exceptional conditions were freewheeling from any viewpoint, and it was a perfect fit for me.

A few one-semester courses taught by regular professors covered scattered specialized topics, and "conference cycles" (ten lectures or fewer) were given by all kinds of short-term visiting professors. There were limited openings for holders of the Ph.D., hence a fear of over-supply, a small number of candidates (called *thésards*), and no justification for investing in a proper graduate school.

Also, the Paris doctorate came in several flavors, because in the past different political pressures had called for different diplomas with anything but obvious titles. The Doctorat de l'Université de Paris sounded splendid, but requirements were left to the discretion of the faculty, and it had no legal value. It was tailored to foreign students who did poorly and might become hostile to France if sent back without some piece of paper. To be short-listed for a position, the only flavor that mattered was analogous to the German (and now also French) *habilitation*, the Doctorat d'État ès Sciences.

For the thesis, I was largely left to myself, a widespread and disorderly practice. Many *thésards* and professors bemoaned it, but for me, disorder was a godsend. Serious teachers and enlightened guides might have done far more harm than good. The orderly United States might have constricted me beyond reason.

Life-Altering Verbal Lashing from Szolem

From the time I quit the École Normale, Szolem had been increasingly bothered by my ways. One day, when I was twenty-eight and stopped by that lifesaving country house near Tulle, he lost his temper, and a polite conversation shifted abruptly to a ferocious verbal lashing, an old-fashioned "visit behind the woodshed."

> *"You are like a student I had before the war, reading everything and always ready to discuss a new book or article. I told him that the next time I saw him in the library, I would suspend his scholarship and let him starve. He took it to heart and wrote a beautiful thesis in no time. . . . It's a tragedy that he vanished during the war.*
>
> *"Too many good students are nothing but well-trained monkeys; they know everything they are taught—and nothing more. If you continue to be of that breed, you will become—at best—a slavish scholar . . . like too many in our family. You can do better. If you want to amount to anything, hurry up and find out what you can do. Settle down—now!"*

Szolem's wife, Gladys, so sweet and almost always close by, repeated the same thoughts more kindly: "You must already have some kind of an idea for a thesis. Try to write it down and see."

Oddly, this episode worked. It literally turned my worldview around—for a while anyway. I ceased to be a know-it-all intellectual dandy and plunged into a serious search for a doctoral dissertation topic.

Gladys convinced me to think of what I had at hand as a possible topic, and Szolem soured me on "well-trained monkeys." Unlike

Szolem, I enjoy intellectual fencing and occasionally showing off. Otherwise—like Szolem—I absolutely stopped having patience for their games.

I do not deny that plain old-fashioned scholarship is a source of enjoyment, including the hunt for old, musty books hidden on hard-to-reach library shelves. Szolem viewed having a quick memory as detrimental to creativity, but in my case it has been neither detrimental nor an empty distraction. Also, the Keplerian style of research that I came to practice happens to be powerfully assisted by flipping through reference books and forgotten texts. The goal is not to copy them passively into one's memory but to link them to one another over high intellectual walls or across wide intellectual abysses. My memory has been a key asset—so far.

A Flawed Ph.D. Dissertation Well Ahead of Its Time

The "woodshed" episode made me listen to Szolem more than usual. For a thesis topic, he suggested a theory to which I have already alluded, one originated in the 1910s by the mathematicians Gaston Julia and Pierre Fatou and now called quadratic dynamics. I did my best but soon gave up—much to Szolem's consternation—because the topic seemed hopelessly "stuck" and, perhaps, because I was a young rebel. Only after a deliberation that lasted thirty years did I feel up to facing quadratic dynamics, and I discovered something that became its most recognized icon—the Mandelbrot set.

Instead, I wrote a somewhat strange two-part dissertation for the Doctorat d'État ès Sciences, which was soon overtaken by far better work. But it largely determined the course of my life and—arguably— the work that led to changes in the course of several sciences.

The first part of the dissertation concerned George Kingsley Zipf's universal power law distribution for words. The other part was an incursion into the foundation of an ancient area of physics: generalized statistical thermodynamics. One of my models of word frequencies relied on that second part in a very exotic form. Unfortunately, this mixture was dreadful academic politics. More important, my thoughts in physics were still very much in flux. In fact, they took many more years to become ready for publication. In 1952, this combination was viewed as wild and every onlooker warned me that it would in no way be perceived as natural. The stretch across the abyss between fields was too extreme. In addition, the first part presented a subject that did not yet exist, and my main goal was not to help linguistics become mathematical but to explain Zipf's law.

Why the rush? At a meeting in London, I had been offered a post-doc at MIT. The desire to take off pushed me to cram everything I had at hand into the dissertation. Lacking any advice, my result was unfinished, and I presented it in a grossly incompetent style.

Any half-friendly professor would have rejected my topic. But I had no Ph.D. adviser. All that mattered was finding a well-disposed chairman powerful enough to place me on the short list of pre-screened candidates for an academic job. Even if my choice had been less exotic, the selection of advisers was pitifully small because the science faculty at the University of Paris had very few professors. Every Ph.D. dissertation had to be printed, and the page facing the title page had to name all the professors, irrespective of discipline. An amazingly small list! Even worse, the few professors at Carva and the Collège de France were denied the right to supervise Ph.D.'s.

Having an actual adviser was a novelty at that time in Paris. Szolem had not had one; only after Szolem's defense did Hadamard read his thesis and become his patron. Szolem told me of a case when he saw Hadamard livid that someone had asked him for a thesis topic and supervision. "Can you imagine that? If he has no topic of his own, he should not even think of a Ph.D.!"

A report had to be written, and a Ph.D. committee had to be selected. By default, the task fell to the sitting professor of probability theory and mathematical physics. That chair's prestige had climaxed with the great Henri Poincaré. Paul Lévy had amply deserved and desperately wanted it, but the Sorbonne faculty first chose the miscast Maurice Fréchet (1878–1973), and then a scientific lightweight, savvy statistician named Georges Darmois (1888–1960), who was moonlighting as the manager of his wife's iron foundry.

Darmois was by nature unfriendly, and we always talked standing in the corridor with many others waiting around for their turn. Taking on an additional Ph.D. student demanded little time and made him look good. He probably took it for granted that I would continue at Philips of Holland and therefore merely glanced at my thesis while on an airplane, having decided in advance—without telling me—that I would only make the meaningless long list of candidates for a job, not the desirable short list.

In any event, how should this dissertation be pigeonholed? The science faculty had no formal departments, and Darmois's chair overlapped mathematics and physics. I could choose either, with inescapable consequences. But my plight interested nobody, least of all Szolem.

Luckily, I chanced to cross paths on the street with the physicist Alfred Kastler (1902–84), a close friend of Szolem and an exceptionally nice man, whom I had met when I was twelve. Later in life, after receiving the Nobel, he proclaimed that a lifelong collaborator deserved equal honor. Splitting the medal was not possible, but he gave Szolem half of the money. Later, he published a book with a French title that translates as *Poems in German by a French European*. Indeed, he was born in Alsace, and until he entered the École Normale on a special deal, spoke only German. During the war, he never yielded to Hitler, kept Szolem's apartment safe by living there, and left as soon as Szolem came back. This man was attuned to nuance and comfortable with living between two cultures. The perfect man to consult. We stopped to chat. I sketched my thesis and described my quandary.

He sighed with foreboding and confirmed that one must not combine two very different topics—especially when neither would lead to a job. Thermodynamics was inactive and jobs no longer existed; quantitative linguistics did not yet exist. In physics, I would compete with a relative flood of strong dissertations on currently fashionable topics. Kastler realized that fate was on my side. Mathematics offered one big advantage. The level of abstraction of nearly all the new dissertations had become so extreme that Kastler and fellow physicists had seen enough. Some new openings were to be reserved for applied mathematics, and—miracle—my chances of landing a job might in fact be rather good.

Real fern

Fractal fern using an L-system

Real clouds

Fractal clouds

Real coastline

Fractal coastline

Real or fractal?
Imitation: the first step to understanding

Surface
dimension 2.15

Surface
dimension 2.5

Surface
dimension 2.8

Fractal forgeries showing the relationship between
fractal dimensions and roughness

"Cool Afternoon"

"Lethe"

Artistic renderings of fractal landscapes

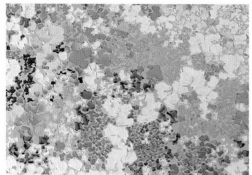

Fractal painting of flowers,
Augusto Giacometti

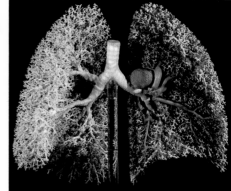

Cast of a human lung

The Great Wave,
Hokusai

Rough deposit of gold

Turbulence on Jupiter

Science, Art, and Nature

Zooming into the Mandelbrot set

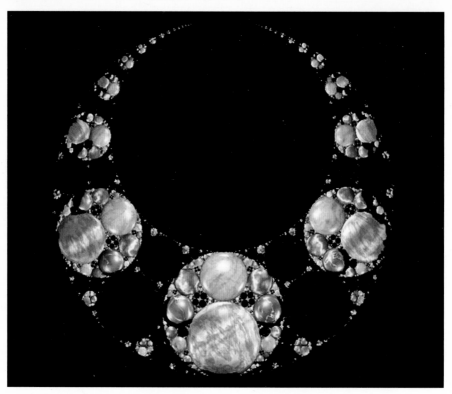

"Pharoah's Breastplate," limit set of circle inversions

Variation of the Mandelbrot set

Deep into the Mandelbrot set

"Cave painting," modified Mandelbrot set fragment

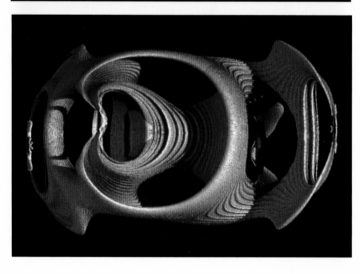

Quarternion Julia sets

This advice proved wise beyond any short-term consideration of bureaucratic politics. In order to accommodate my longings around 1950, and also my life's accomplishments in physics or mathematics, the scope of either science must be given a very broad interpretation. It used to be that terms like "mathematics" were defined broadly, and physics actually began as a corner of mathematics. A hundred years ago, however, it separated from mathematics and from engineering. Only recently has physics expanded again, both in directions where it becomes hard to distinguish from mathematics and in directions (hard and soft materials) where it becomes hard to distinguish from engineering. For example, several Nobel Prizes have recently been granted in physics for work that in decades past might not have qualified. Back in 1950, when it had a big impact on my life, Kastler was right that my work belonged in broadly understood mathematics.

Darmois agreed, but decided that my committee's chairman need not be a mathematician. His unexpected choice was Prince Louis de Broglie (1892–1987)—the official reasons being that de Broglie publicly praised interdisciplinary work and that my thesis would benefit from association with a broad-minded professor familiar with flying solo. Twenty-five years earlier, this aristocrat had been a key contributor to quantum theory.

The cover of every French dissertation of that day referred to a second thesis, with the universal title *Propositions données par la Faculté*, which was never published. This all-purpose boilerplate denoted a reading course requirement meant to amplify and balance the skimpy graduate course work. By unwritten tradition, the topic had to be very different from the main one. A heavily computational first (real) thesis could be balanced by assigning a philosophical issue.

My assignment was long and heavily computational: the then-recent Ph.D. dissertation of the mathematician Yvonne Choquet Bruhat (1923–) asked a key question: Do the equations of gravitation discovered by Albert Einstein have a solution and only one? Physicists found this question of no interest, but mathematicians found it very difficult and hence fascinating. Bruhat managed to prove that it was sufficient for the initial conditions to have well-behaved derivatives at least to the magic order of 7.

At my thesis defense, I was reporting on this proof elegantly enough when Darmois suddenly broke in. "Your presentation was excellent. But could you tell us more specifically why the topic of your second thesis is important?" The second thesis was a report on a very long recent article on gravitation—a fine but early and provisional technical stage in a long-range program that was bound to (and did) proceed much further. As I was fumbling for a suitably noncommittal response, Darmois smoothly took over. He turned ninety degrees to address face-to-face the committee chairman—who was none other than de Broglie. We were informed that something Darmois had written in the 1920s on relativity theory deserved mention.

At that point, it became limpidly clear why Darmois agreed to report on my thesis topics and suggested the topic he did for the second. He was campaigning for election to the Académie des Sciences as an astronomer—his early area of expertise. So he welcomed an opportunity to give an uninterrupted twenty-minute presentation to the academy's perpetual secretary, de Broglie. This political maneuver soon succeeded.

At no time did I feel that any member of my Ph.D. committee gave serious thought to the content of my thesis. Since then, my experience with these committees has made me realize that mine faced an impossible task. An unwise and dreadfully rushed presentation did not help, and my excuse—a waiting postdoc in the United States—was weak. Even a flawless job would not have affected the fundamental obstacle noted by Kastler: my key topic was far from any mainstream. At the time, even I did not know that my dissertation was but a seed from which a mighty tree was to rise.

Do I Regret This Messy Doctorate?

I do not regret my messy doctorate in the least. It was preferable to falling under the supervision of a maniac who would delay me until I had been shaped to fit his own agenda. On several occasions in my life, an element of freewheeling in the system proved a blessing.

The irony of my failing to be short-listed in 1952 is that it no longer mattered by 1956. Enrollments were exploding, jobs were opening everywhere in France, and every passed-over warm body languishing in a dark corner was short-listed. Darmois telephoned (!) to inform me that I was urgently needed; in fact, he allowed me the luxury of selecting Lille. I could live in Paris, with a commute of only two hours.

Recounting those long-past events never fails to both amuse and hurt. The choice of a Ph.D. label was one of many critical decisions I faced in my life with no precedent to help. In each instance, a wrong choice might have thrown my life orbit in a totally different and possibly very unfortunate direction. Furthermore, those critical choices had steadily increasing consequences. As a result, the great promise I had held at age twenty had largely dissipated. In between, I suffered many bruises and not a few indignities. But in hindsight, I was the only one to blame, or to praise, since my lowly dissertation turned out to be the germ of all I went on to accomplish.

Fortunately for me, my corner of French education had little faith in classroom teaching and great faith in its ability to select the best among self-motivated individuals. If a dissertation is to be written solo, no special arrangements are needed. Among bad alternatives, an incoherent or indifferent environment is best.

Take my Ph.D. chairman, de Broglie. *His* dissertation—which became one of the two sources of quantum mechanics—was also written with no help. Fifty years after it was defended, the question arose: How did it manage to be accepted? A committee member who was still alive confirmed the rumor that the great physicist Paul Langevin had found the thesis incomprehensible. But he could imagine no harm in accepting it because de Broglie would surely follow the example of his elder brother, Duc Maurice de Broglie, and never apply for a job. However, he sent the thesis to Einstein, and the rest is history.

The LEP Laboratory of Philips Electronics

While I was a Ph.D. student, the graduate fellowship from the National Center for Scientific Research (CNRS) would have been extraordinarily meager. I preferred to lead a parallel life. Having finally satisfied the air force, I first followed inertia; that is, I simply resumed the line of thought that had already led me to aeronautics at Caltech in 1947 and sought something on the interface of mathematics and flight. At ONERA (the French counterpart to NASA), a potential supervisor confessed that he had not enough ideas for himself and could not supervise me. Nevertheless, I met the proper authorities, interviewers expressed great enthusiasm for my qualifications, and I was introduced to the big boss. He assured me that his official approval would come in a few days, but he also held other jobs, micromanaged everything, and clearly spread himself too thin. Days and weeks passed with no letter, only telephoned reassurances that the paperwork was on the boss's desk waiting to be signed. Father's chronic worry proved true, and he saw additional evidence of how dreadful it was to be employed by a state agency. Aeronautics had become less and less attractive to him. Quietly, he started scanning the newspapers for openings for technical positions.

One job listing that he really liked gave an address but no name. He found out that it was from Philips SA, the closely held French branch of the Dutch-based electronics multinational. More precisely, it came from a new research division, LEP. Not the obvious Laboratoire d'Electronique Philips, but Laboratoire d'Électronique et Physique Appliquées. Father saw a seamless move from aeronautics to electronics within a big international company. Even if a revolution broke out in one country, they would reassign you to some other subsidiary.

Philips was seeking an alumnus of a *grande école* fluent in English and well informed of a technique called spectral analysis. I was a *polytechnicien* with very high rank, had spent two years at illustrious Caltech with top grades, and knew about spectra in two ways—Caltech friends made practical use of them in turbulence research, and I had "inherited" elements of the theory from Szolem.

The job was a perfect fit for both sides. Within days, I was called by the Dutchman who ran Philips France from the elegant avenue Montaigne near the Champs-Élysées. Soon came a signed and stamped job offer, with a salary higher than the one ONERA had promised.

Unlike Father, I mostly favored Philips because it seemed compatible with writing or even inspiring a doctoral thesis. But what made

Philips so interested in spectra? The hope was that I would fulfill a genuine and basic need. Reluctantly but feverishly, the TV industry was then preparing for color, and it faced a little-known technical quirk—letting white light go through a prism. Newton had analyzed white light into a "spectral" image made of colors ranging from red to violet. Similarly, a sound can be analyzed into pure sounds of all frequencies, and in the simplest example, into a fundamental and its harmonics. Hence the alternative terms "spectral" and "harmonic" analysis.

This analysis inspired a color TV system, perfected by RCA and GE engineers, called NTSC, after the National Television Standards Committee. It's now ancient and obsolete but—like QWERTY type-writers—still in use. As they were learning the ropes, the engineers at the French subsidiary of Philips needed a theoretician to hold their hand.

Several countries in Europe developed better systems. Poor old NTSC came to be reinterpreted as meaning "Never the Same Color." To ensure compatibility with the existing black-and-white receivers, the signal included a detailed image that old TVs would interpret as black but color TVs would interpret as green. The full-color image was achieved by adding red and blue images that were much fuzzier than the green one.

Philips colleagues were struggling to improve on the black-and-white iconoscope by designing and then building an exquisitely complicated device called a supericonoscope. Failure upon failure—then success. Shortly after those engineers had achieved consistent success, they were transferred to a distant factory and began mass-producing that contraption.

My role model at Philips was the distinguished physicist Hendrik Casimir (1909–2000). He was the technical director of the Philips Research Laboratories in Eindhoven—arguably close in quality to the famed Bell Laboratories. On a dime, he could turn between science, technology, management, and corporate or national policy, and he instantly got the point of every presentation. His visits to check on us were irregular but rumored not to be random. Eindhoven was a provincial one-company city, and educated Dutchmen of his time were fluent in French, so successful new shows invariably brought Casimir back to combine business with pleasure in Paris.

Philips sported an extraordinarily revealing history. The family of the founder, as it happens—hard to believe—were close relatives of Karl Marx! They owned a tannery in Eindhoven, so the nascent company enjoyed low labor costs without having to move. The Netherlands took until after 1900 to sign the international agreement on copyrights—as revealed by those old French books that I read as a child in Paris. The same was the case for patents.

After postdocs at MIT and the Institute for Advanced Study, I found that Philips no longer had use for me. The TV group had advanced from research to development. My stint at Philips was short, but I learned many things. In a way, working for industry was a rehearsal for a far longer stint at IBM, and working experience with spectra was very useful indeed.

Father Dies in 1951

Shepherding me toward Philips became the last in the long series of gifts from Father. I don't recall either him or Mother being sick in bed or visiting a doctor. Their explanation was that less fortunate persons would have perished early in one of the catastrophes that they had managed to sail through.

But cancer struck. First a successful kidney operation brought several years of good enough health. Then lung cancer struck. Father refused a second operation, and his doctor advised heavy doses of radiation, the outcome of which would be quick—either way. We later found that every encyclopedia in the house had bookmarks under "cancer." Besides, every day the newspaper brought a detailed update on fellow sufferer King George VI of Great Britain.

When my first paper came out, Father was so ill that I could not wait for reprints to be made, so I borrowed a library copy. It was not clear if he fully understood what I was showing him. He died a short time later.

Chronic lack of money, hopeless overwork, and Father's exhausting business travel—and then illness—meant that my parents could rarely entertain. So we expected few mourners, but a small crowd joined Mother and her two sons at Father's burial. Szolem had been traveling, and the ceremony was held up until he came back. Before leaving, he stopped by and ended by saying, "*À bientôt donc.*" With all of Szolem's travels, we wondered if this "See you soon" was meant to be understood as "if fate allows."

Unexpectedly, Mother insisted on a religious funeral. The rabbi in Brive during the war—who surely had sent that Angel to protect us—had conveniently moved to Paris, was found, and agreed to officiate. His eulogy was neither corny nor canned. He recalled, with surprisingly many personal details and in very warm terms, that though the war had led him to meet many parents ready for any personal sacrifice for the sake of their children, none had come close to Father.

14

First Kepler Moment: The Zipf-Mandelbrot Distribution of Word Frequencies, 1951

"TAKE THIS REPRINT. That's the kind of silly stuff only you can like."

These words of Uncle Szolem—ending a visit—opened a door. The words in this reprint first seemed narrow and undistinguished, then profoundly flawed. But I figured out how to correct this flaw and—endless surprise—in the hour that followed experienced my first Kepler moment.

I allowed my finger to be touched by a complicated set of gears that soon grabbed my body—and never let go. In a different analogy, I found myself in the position of that child in a story who noticed a bit of string and—out of curiosity—pulled on it to discover that it was just the tip of a very long and increasingly thick string . . . and kept bringing out wonders beyond reckoning.

Oddly but almost ineluctably, that string, that reprint, ended up directing me to some of the main themes of my scientific life: unevenness, inequality, roughness, and the concept of (as well as the word) fractality. On many occasions, I was to feel that the topic was nearly exhausted, that little else remained to be said—but it kept reappearing from a totally unexpected direction.

A Fateful Metro Ride

At the end of a day spent near the Sorbonne, it was not much of a detour—before taking the metro home—to stop at Szolem's flat. A chat in his study often turned to debate.

The quote opening this chapter was Szolem's response to my routine request for reading material for the long ride home. That day, he pulled out of his wastebasket a reprint he had recently received from the Harvard mathematician Joseph L. Walsh (1895–1973), president of the American Mathematical Society. This reprint was a friendly review in the popular monthly magazine *Scientific American* of a book titled *Human Behavior and the Principle of Least Effort*, written by George Kingsley Zipf (1902–50). Independently wealthy, this academic character was a university-wide lecturer at Harvard in a self-invented field he called statistical human ecology. His topic was the oddest imaginable—an absurdly simple mathematical formula that claimed to be a universally valid summary of a mass of empirical observations on how the words in ordinary writing are distributed between common and rare.

I became hooked: first deeply mystified, next totally incredulous, and then hopelessly smitten . . . to this day. I saw right away that, as stated, Zipf's formula could not conceivably be exact. But the metro ride was long and I had nothing else to do. By its end, I had derived a more general version I could explain and was dying to confront it with data. I soon decided to pursue this strange avenue, all the way to a Ph.D. It is known today as the Zipf-Mandelbrot law.

Everybody—above all, Szolem and Marcel-Paul "Marco" Schützenberger (1920–96), a man I had recently befriended—was aghast. They saw Zipf as a crank! Counting words was neither real math, nor real science, nor real anything. Nobody with even minimal technical skill was interested. It would never lead to a proper job. No professorship. Marco located Zipf's book and made me take a look. On the whole, it was indeed dreadful. But if you could ignore the text and believe the graphs, they covered many fields and were fascinating. They contradicted Zipf's claim about word frequencies that Walsh had accepted—but confirmed the Zipf-Mandelbrot formula-to-be!

So I could respond to my friends by broadening Plutarch's advice: to admire part of a man's works, you need not admire everything the man claimed. To my rational side, the fact that science's central casting office considered Zipf an oddball was not sufficient reason to disregard him. To my herd-averse, rebellious side, it may even have been a plus.

In short time, the Zipf-Mandelbrot formula became part of my Ph.D. dissertation. Then other graphs in Zipf's book filled several years with interesting developments. I then left Zipf behind and allowed my path to be guided by logical necessity, pure chance, or unabashed play. Eventually, all of that led to fractals.

Inequality and Unevenness Are Everywhere

How long does a book on the best-seller list remain there? Most stay for a few weeks, but a few may remain for a hundred weeks or even more. This extreme inequality is basic publishers' folklore.

Type a name into an Internet search engine. Some names draw a blank, many have a small number of hits, but a few draw millions of hits. Think about the geographical areas of islands. Greenland and Madagascar are huge, while a countless number are tiny. What about the inequality of sizes of the states in the United States? What about the even greater inequality of areas of the French provinces before the Revolution cut them into near-equal departments, of Soviet republics as contrived by Stalin, or of the parts of present-day Russia?

Extreme inequality is a familiar pattern in nature and in the works of humans. Such distributions are called long-tailed distributions. For them, no value is typical, and the contrast between short and long tails came to play a central role in my work.

Most long-tailed distributions have important consequences, but the papers and books written on this topic over the years were disappointing. My luck was to begin with the distribution of word frequencies—a thoroughly atypical example without any important consequences, and uniquely easy to handle.

Incidentally, in 1952, my first involvement with long tails involved no computers. I first saw a computer in 1953 and first used one in 1958, after I went to IBM.

Zipf's Universal Power Law for Words

In written text or in speech, some words, such as "the" or "this," have a well-defined frequency. Other words are so rarely used that they have no defined frequency. Here was Zipf's game: Pick a text and count how many times each word appears in it. Then give each word a rank: 1 for the most common word, 2 for the second most common word, and so on. Statisticians rarely use this method, but there is nothing wrong with it. Finally, graph the frequency of each word against its rank.

An odd and hard-to-read pattern emerges. The curve does not fall gradually from most common to least common word. At first, it plunges vertiginously, then drops more gradually, continuing in a long tail that declines very slowly—like an exaggerated profile of a ski jumper leaping into space, to land and coast down the gentler slope below. By the very definition of rank, frequency varies inversely with rank. Zipf claimed something far stronger: it is about one-tenth of the inverse of rank. So the product of a word's frequency and its rank is approximately equal to one-tenth. The curve almost merges with the coordinate axes—making it near impossible to read.

To compare such curves, it is best to replot them more legibly by replacing both the rank and the frequency with their logarithms. While it may be a bit scary, this word denotes something quite innocuous. A number's decimal logarithm is roughly its length when it is written using the standard digits from 0 to 9. More precisely, it is smaller—by at most 1—than that number of decimal digits. Thus, the logarithms of numbers from 100 to 1,000 grow from 2 to 3. Taking Zipf's claim that each word's frequency is exactly one-tenth the inverse of its rank, on a doubly logarithmic graph, it follows that the data fall along a straight line with a slope of −1, one that decreases vertically by 1 for every horizontal increase by 1.

The language—English, French, Latin, whatever—does not matter. Neither—quite oddly—does the writer's degree of literacy. This is an example of what physicists were soon to call a universal relationship. Another notion in physics, called scaling, is one that underlies

153

fractals. Zipf, eyeballing his charts and fitting a curve to the data, devised a formula for it. Walsh featured that formula and observed that it baffled everybody who looked at it. Inspecting those graphs cautiously and critically was a practice that physics came to adopt around 1900, then revived in the 1970s and 1980s—and one I have espoused since the early 1950s.

Unfortunately, Zipf's assumption yields conclusions that are simply impossible. For example, it implies that, as a text unfolds, roughly every tenth word has not been used before. One would expect new words to enter at a gradually decreasing rate. Worse: by the definition of frequency, the different words' percentages *must* add up to 100 percent—but Zipf's formula contradicts this absolute mathematical requirement. One facile way out is to "truncate": to assume that new words stop being added as soon as the total number of different words has reached 22,000 (the exponential of 10). How could such a universal limitation apply to both James Joyce and an illiterate? In fancy words used by physicists around 1900, Zipf's original law suffers from a "divergence difficulty" called an "ultraviolet catastrophe," making his claims mathematically self-defeating.

Might this be the reason that everyone who looked closely dismissed the whole silly business? Zipf's claims seemed admirably objective but actually hid the fact that on Zipf's graphs the product of frequency and rank is not the universal constant one-tenth. It varies! However, let me confess that I also did not immediately pay attention. I recall accepting, for the sake of argument, that the original formula represented the data to some degree, and attempted to reduce it to some basic principle—free of any "catastrophe" that might account for James Joyce, illiterates, and others in between.

The fact that it applies to all languages—is universal—implies that Zipf's law is irrelevant to the core of linguistics, which is grammar. In one of the very few clear-cut eureka moments of my life, I saw that it might be deeply linked to information theory and hence to statistical thermodynamics—and became hooked on power law distributions for life. Those "details" had eluded not only Zipf—not trained as a scientist or mathematician—but also Walsh. Anyhow, appreciating the history of ideas does not make a street-smart scientific explorer. My good fortune resided in an unfair advantage. I was to be the first—and for an interminable time, the only—trained mathematical scientist to take Zipf's law seriously.

The Kepler of Word Frequencies?

Why do I view that fateful metro ride as a Kepler moment? For Kepler, the role of toy had originally been played by the ellipse, an esoteric geometric curve with little known application. I dealt with an esoteric wrinkle in the study of language as it stood in 1950. That wrinkle—statistical thermodynamics—is one of the most sublime pillars of physics.

The key feature of the Zipf-Mandelbrot formula exponent was inherited from the statistical thermodynamics motivation: a "temperature of discourse." It could measure differences from text to text, from speaker to speaker. It gave a numerical grade to the richness of someone's vocabulary. Low temperature, limited vocabulary. High temperature, rich vocabulary. The original Zipf's formula is a very close approximation—but misleading. Joyce's *Ulysses* was welcomed by Zipf because it was long, but also because it was atypical. The temperature of discourse could become a powerful tool of social measurement by capturing erudition in a number.

So that long metro ride witnessed the first of many Kepler moments in my life. Soon after it, I examined Zipf's book. His charts confirmed that the Zipf-Mandelbrot formula was a vast improvement. A difficulty: a well-defined probability may exist for common words, but what about rare words, especially in multiauthor works or composite files of newspaper articles? In due time, I identified many problems—which still remain open.

Those events taught me a fundamental lesson—that an applied mathematician's relation to reality is fraught with problems. Worse, experimentalists try to help by simplifying what they see, and key facts are often unwittingly overlooked. They must be respected but never trusted without question.

A Case of Haste Rewarded?

Miraculously, the paper reprint of Walsh's review of Zipf remained in my files and came to light as I was writing this memoir. I made a point of reading it again. It's clear to me now that in my excitement I read it casually and rushed to work. Walsh's review also contained these words I had either missed or forgotten:

> It would be rash to prophesy that a new science of human behavior will now evolve along the lines of the history of mechanics, but it would be foolish to ignore the lessons of that history. . . . Tycho Brahe . . . made numerous observations of the motions of the planets [that were] used by . . . Kepler . . . to formulate . . . fundamental laws . . . and Newton . . . in turn [to found] the science of mechanics. . . . Opportunity is ripe for new Tycho Brahes, Keplers and Newtons! . . .
>
> It might be fruitful to investigate speech as a natural phenomenon . . . a peculiar form of behavior . . . in the manner of the exact sciences.

Shame on me! I had forgotten that Walsh mentioned Kepler by name. My first Kepler moment concerned long tails, an uncanny fit to my wild dream. I had to be reminded of these words. Yet I recall being spellbound by the Keplerian possibilities and not bothered by the absence of geometry—which became central to my work.

Early on, a shadow was present—the example I worked on was devoid of important consequences. No one could predict that I was to be called the "Kepler of word frequencies," then more generally the "father of long tails." In a fifty-year time span, they went from an aberration hardly worth mentioning to the center of wide attention in the early 2000s. Had I approached it from a seemingly more "worthy" angle, I am convinced I would have failed. My luck was holding.

Questions rush in. Computer searches reveal that Zipf was reviewed favorably and—among the mathematically unsophisticated—had found a following. Had Walsh noticed that Zipf's original formula was nonsense, he would have discouraged his friend. Why did that review fail to attract the attention of anyone else of at least adequate mathematical competence?

From Unruly Beginner to "Father of Long Tails"

I had in hand the key topic of my doctoral thesis: the very simple mathematics behind the unexpected distribution of word frequencies. On December 19, 1952, the die was cast. My Ph.D. dissertation loudly affirmed a Keplerian determination to become a solo scientist—the kind my world thought had vanished. Figuratively, I was choosing to be an apprentice-hermit at a time when science was rushing to adopt the ways of the more structured religious churches. Convinced that this direction could never be reversed, I stopped thinking of ever contributing to plain mathematics or physics. Well . . . I eventually did— very late in life and with a vengeance.

While carefully thought through, my dissertation was "technically" easy and imperfectly written. It barely matched my ambition— but I was in a rush and underestimated myself. Plain old-fashioned luck, and perhaps a learned skill of turning difficulties to assets, made me the first—and for a long time the only—mathematically competent person to face long tails squarely.

Determination but No Foresight

I simply loved being able to do everything all by myself. With my records at Normale and Carva still widely known, my thesis was approved when it was clear that nobody in Paris much cared about my topic or my career. Of course, my parents, Szolem, and many others pushed in conflicting ways. So—for better or worse—I did it in my own way.

I did it with determination but no foresight. How should I follow up? Szolem had warned me in no uncertain terms that—before rushing to Caltech—I should identify in Paris a suitable combination of a topic and an adviser who could protect me for a while. Otherwise, nobody would help me find a job. I was beginning to wonder about my chances for a proper academic career in any country.

Against the background of an early life of hard knocks of every kind, was I acting like a spoiled child? I had a free built-in insurance. Not only was I still a member of the CNRS, but the Ph.D. had earned me a promotion. More than a few of my contemporaries stayed at the CNRS, kept quiet, and pursued activities they carefully failed to report. So, no, I was not acting spoiled. I did not want to hide—I wanted to find the best conditions to fulfill my Keplerian dream. Dreams can be burdensome.

My political innocence would not be punished. In a few years, explosive growth overwhelmed the old French universities, and many new lifetime jobs had to be created. That Ph.D. certificate would come to matter greatly. But in 1952, this growth was but a distant hope. I knew that a hastily written doctoral thesis in French and devotion to a field that did not officially exist were unlikely to be enough. In any event, finding companions and putting down professional roots demanded some commitment.

Luckily, short-term jobs were plentiful—though mostly not in France. During the five years after my Ph.D., I sampled several thoroughly different ones, making my life extremely interesting and varied. But, like at Caltech, I didn't accomplish much new. Eventually, inspiration did come from my Carva teachers Paul Lévy and Gaston Julia—but only through their work. Had I sought their advice, I am sure I would not have taken it.

15

Postdoctoral Grand Tour
Begins at MIT, 1953

I RECALLED THOSE WORDS *Gaudeamus igitur, juvenes dum sumus*—"While we are young, let us rejoice." In my case, rejoicing did not mean carousing. It first meant a highly unorthodox Ph.D. boldly asserting the kind of work I hoped to carry on. Later, it meant a modern form of a different medieval tradition: that of apprentice-scholars' wandering years, which I think of as my "postdoctoral grand tour." During that time, I worked near the two exalted living role models to whom my thesis was dedicated; mathematicians of the highest rank, they had repeatedly achieved the Keplerian dream I wanted to emulate.

The first was Norbert Wiener, a professor at MIT, the Massachusetts Institute of Technology, in Cambridge. He had authored an unusual book I greatly admired: *Cybernetics, or Control and Communication in the Animal and the Machine.* "Cybernetics" was a word Wiener had just coined, and the title defined that word as ranging from brains to telephone switchboards.

The second was John von Neumann, a professor at IAS, the Institute for Advanced Study, in Princeton. After MIT, I became von Neumann's last postdoc there. He had written, with Oskar Morgenstern, *Theory of Games and Economic Behavior.* Both titles promised new frontiers and new topics—or at least altogether new combinations of existing topics.

My Ph.D. dissertation's title, *Games of Communication*, overstressed a bit my devotion to both men—whom I perceived as made of stardust. These two men were the only living proof that my Keplerian dream was not an idle one—that it was possible to put together and develop a new mathematical approach to a very old, very concrete problem that overlapped several disciplines. Matching the sterling quality of their accomplishments was far beyond my ambitions, and I couldn't think of less exalted advisers.

Norbert Wiener of MIT

The towering Keplerian achievements of Norbert Wiener (1894–1964) were his mathematical theory of Brownian motion and cybernetics—the word and the book. Isaac Newton knew around 1700 that prisms decompose light into components of different colors. But the mathematical theory was given much later, by Wiener. A related achievement, his theory of Brownian motion, strongly affected me later in my life—as a miserable model of the variation of competitive prices, and as a wiggle with an interesting boundary that forms fractal islands. His own account of early motivations was thrilling. Having become interested in the motion of pollen as seen through a microscope, he decided that the solution must use something called Lebesgue integrals—at that time still novel and the epitome of an esoteric toy.

As Wiener's follower, I never tried to further develop the technical problems he had raised. I preferred to either move sideways and open new technical problems or take conceptual new steps by going beyond the Brownian realm. Yet Wiener's work has remained a shining beacon for me.

He was a mathematical genius—a widely celebrated establishment figure. He became the leader of a scientific avant-garde that he hoped would grow to cover communication and control in machines and living things. To denote this goal before it was even partially fulfilled, he drew on a Greek word to coin "cybernetics." I heard this word early on. When Wiener was in Paris in 1947, Szolem invited him for lunch and asked me to join them for coffee.

He was a master in a field of mathematics very close to Szolem's, and they had written joint papers. Szolem looked up to his barely older friend, but acknowledged that his and Wiener's mathematics had been born in concrete contexts that had occurred centuries ago. Fresh inputs from science were, for Szolem, intolerable, and there was always an undercurrent of irritation.

To a mathematician, the term "function" often denotes something that varies in time. Wiener preferred to use "noise." Szolem was bothered and wondered aloud. Was this a mannerism left over from consulting for the military, or a way of showing off? I argued that Wiener's esoteric mathematics was genuine, part of a lifelong ambition to understand physical fluctuations. He wanted to "see over the fence" to engineering, biology, and social sciences—but not to narrowly defined economics.

Jerry Wiesner's RLE, an Ideal Research Environment

Fired up by Norbert Wiener's cybernetics . . . that unique scientific incubator, the Research Laboratory of Electronics (RLE), has for two decades provided an almost ideal research environment and has been a model for the structure of other research centers. . . . [Back in 1946,] We could hardly imagine the excitement and intellectual pleasure that lay ahead of us. In fact, as I look back, I have the impression of powerful personalities and even more powerful ideas drawing people together from all over the world. My memory is a great pleasant blur, not unlike my mental movie of the spontaneous creation of the universe.

These words were spoken by Jerry Wiesner at MIT during the twenty-fifth anniversary celebration of RLE, a most remarkable institution, where I went after my doctorate to continue my education. Jerome B. Wiesner was Professor Wiesner when I met him as head of RLE, later became Dr. Wiesner, and ended his career at MIT as the unforgettable President Wiesner. We knew him as Jerry.

In this quote, the words "from all over the world" are essential, and the word "spontaneous" toward the end is very important. All those who knew Jerry can testify that in his "mental movie" he did not view himself as a creator but as a facilitator. In fact, he was the rare manager who could make creation seem to occur spontaneously. RLE was a remarkable hybrid of solo scientists of the ancient academic tradition and the more modern group of academics inherited from MIT's famed wartime Radiation Laboratory, where radar had been developed.

This was the heyday of RLE. Jerry was almost the complete opposite of Wiener, though the similar names (especially when enhanced by foreign accents) often led to—mostly innocuous—confusion.

For a big boss at the Radiation Laboratory during World War II, Jerry Wiesner had been astonishingly young. He was not himself an accomplished scientist, but was endowed in an unusual way: he had a keen eye for full personal commitment (a large part of scientific value), a highly developed sense of noblesse oblige, and the ability to interact with everyone and get things done. He simply knew how to run an organization without self-aggrandizement, with invisible bureaucracy, and with maximum respect for his charges—including many thoroughly spoiled brats. Next to him, I always felt like a child.

Early on, Jerry became close to Senator John F. Kennedy. After JFK ascended to the presidency, he took Jerry as his science adviser—more effective and visible than those who came before or after. Back at MIT, Jerry moved up by stages and became its president during a period when the New Left was riding high and the institute actually appeared endangered.

By chance, Jerry heard me lecture in London in 1952 and liked my talk and the fun discussion it provoked. You must believe it—I argued with the ethnologist Margaret Mead (1901–78), who had gained fame studying sex in the South Seas! That she attended my lecture illustrates the wonderful open mood of those distant years. As was his style, Jerry invited me over to MIT with practically no paperwork. He was then an associate professor and, like several other staff, sat in an open cubicle in a large room. This kept him close to the troops.

RLE was housed in the large and labyrinthine Building 20, a quickie wood, tar, and asbestos barracks that the Radiation Laboratory could perpetually adjust to changing needs. Like everything else, my chair was beat-up and shaky, but high-class—a sign affixed on its back read LEE DUBRIDGE, former big boss of Rad Lab and Caltech president in my student days. That chair's survival testified that when freewheeling scientific research is properly managed, it is not a financial extravagance but a true bargain.

Northeast of what was Building 20, a neighborhood called East Cambridge is now filled with high-rise industrial labs and upscale housing—where I live. At that time, it was a low-rise mix of industries and tenements. Therefore, RLE was often filled with either the aroma of a traditional chocolate factory or the stench of a rendering plant that boiled carrion into pure white soap. I took all this as constant confirmation that the process of creation is intrinsically messy and suffers more from soulless order than from surrounding physical decay.

Controversial Balance Between Conjecture and Proof

Claude Shannon (1916–2001) was the intellectual leader whose wartime work, published in 1948, created information theory and provided RLE with an intellectual backbone. His work on noiseless channels was a point of departure for the theory of word frequencies presented in my Ph.D. thesis.

But far more impressive was his noisy channel theorem. Actually, it was not a theorem at all, only a brilliant conjecture—in a style that is controversial, and of which I eventually became a very active supplier. The point? Even an arbitrarily noisy channel may be programmed in such a way as to allow it to transmit messages with an accuracy as close to perfection as desired.

Shannon's conjecture was plainly an important event, but his proofs were incomplete. Adding the years during which his work was classified, an increasingly clear and general proof was slow to come. The information theorists perceived this as a minor annoyance. But the mathematicians thumbed their noses, noting that Shannon's noisy channel theorem was unproven.

On a later visit to MIT, I played a role in the first proof of this theorem when Amiel Feinstein, a graduate student in physics, came to see me. He was seeking a new topic in electrical engineering that would promise a quick Ph.D. Momentarily irritated by his arrogance, I blurted out that he might try to prove Shannon's bold claim. I explained the issue, mentioned several others who had tried very hard and failed miserably, and wished him good luck. He soon came back with a proof! Stylistically, it was relentlessly pure mathematics, written with no supervision, by a raw apprentice. His proof was checked, found to be correct, and—once touched up a bit—earned him a Ph.D. in physics. But he received little recognition and soon dropped out of scientific competition. The bulk of the credit stayed with Shannon. This was fair.

Noam Chomsky and László Tisza

My fondest recollections of RLE are of a field one would not expect to have found at the industrial MIT and gritty Building 20. Claude Lévi-Strauss, the illustrious anthropologist I had worked with in Paris, had recommended me to his close friend, the linguist Roman Jakobson. Next I met a Harvard junior fellow, Noam Chomsky, and learned about his project for the future of linguistics. In 1953, it was a wild dream, worlds away from existing mainstreams. Along with many others, I wondered whether and where the new linguistics could find a shelter to survive and develop. Chomsky's extreme and often restated positions on broad political issues decreased the odds. To his credit, Jerry Wiesner arranged a home for linguistics at the least likely place, MIT. Chomsky stayed and rose to be Institute Professor. The time I spent playing with linguists was wonderful and educational, and left many lasting friends. Roman Jakobson wanted me to forsake seeking new Kepler thrills and make my home in linguistics. But the more I watched, the clearer it became that linguistics was to be dominated by Chomsky. I soon convinced him and his followers of one big thing: Zipf's law was the basis of an important physics-like (thermodynamical) aspect of discourse, while grammar is like the chemistry or algebra of language. As planned—but not until I had received a first small serving of the combined effects of being admired, I found the field disappointing and moved on.

Except in cases of extraordinary longevity, friendship with an older colleague is generally brief. Thus, it was a rare privilege that my friendship with the physicist László Tisza (1907–2009) lasted far longer than usual. He obliged by being born on 7/7/07 and on the next 7/7/07 provided me with my only chance so far to talk to someone in the process of turning a hundred years old.

The man was short, slight, retiring, and soft-spoken. Upon meeting him, I was told that he had been a well-known and productive researcher—in fact, had come close to fame by almost explaining a curious phenomenon called superfluidity of very low-temperature helium. However—as was added immediately—serious mistakes in that work had to be corrected by his onetime adviser, a star physicist named Lev Landau (1908–68). In truth, Tisza had made no mistake and deserved the credit he did not receive. Tisza was victimized by Landau, but lived long enough for this to be recognized. Instead of clamoring for full credit, he nominated Landau for a prize for this work.

Tisza and I interacted intensely for a few years after a symposium on information theory held at MIT in the summer of 1956. The paper I presented there described an axiomatic for statistical thermodynamics that developed from the second half of my Ph.D. thesis. Asked to comment on my advance text, Tisza praised it handsomely and described himself, on this occasion, as being my student! Given the age difference, his words were a rarity—balm on my heart.

Tisza was a hugely helpful professor. I was delighted to trigger an early celebration of his centennial. A large room was filled, a few people came from far away, and the mood was warm and altogether cheerful. His life had produced little needless sound and fury addressed to outsiders, and much reflection for his friends and his own pleasure. It extended late and added at least one solid brick to the permanent edifice of physics. Many mysteries remain open, but long live diversity. I was very moved.

Effects of Prosperity on the Sciences

Why did I find RLE so attractive? Because it was close to Wiener, but mostly because of its ambition to be the kind of place described earlier. Nearly isolated in Paris, I was eager for a more open and varied environment to live in and to help me decide whether to continue in the direction of my Ph.D. thesis or move on.

I admired this great incubator of imaginative science and engineering and was disappointed that the format I had known did not last. But there were many reasons why it could not. Like my Ph.D. topic, RLE's seemingly perfect timing had arisen not from brilliant long-range planning but from a postwar period of trust in the benevolent power of science and a buildup of expectations that relied on many outside factors. The spiritual health of RLE greatly depended on the financial health of the communications industry. As technology moved on, the role of incubator passed to computers—and so to different institutions, such as IBM Research.

Early on, outside pressures imposed a certain degree of breadth and cohesion on university departments to create overlap. But when mathematics and physics became suddenly rich—when a rising tide lifted all boats—they indulged in a sort of "ethnic cleansing" and restricted their scope down to very pure, or core, topics.

In sharp and most fortunate contrast, communications, and later computers, were atypical. Each chose to interpret their role in very broad fashion. Sadly, RLE's miraculous mix of old and new academic technology and science is only remembered by a few old men.

16

Princeton: John von Neumann's Last Postdoc, 1953–54

"I MUST PROTEST! This is the worst lecture I ever heard. Not only do I see no relation to the title, but what we have heard makes absolutely no sense at all!"

We were in Princeton at the Institute for Advanced Study (IAS), and a luminary named Otto Neugebauer (1899–1990), a mathematician who had made himself a famous historian of Babylonian astronomy, was commenting on a lecture I had just finished.

I stood frozen with gaping mouth as the physicist J. Robert Oppenheimer, father of the atom bomb, sprung up. "May I respond, Otto? If Dr. Mandelbrot will allow, I would like to make a few comments. The title listed in the announcement of this lecture was tentative and should have been changed. But I had the privilege of hearing about his work. I am impressed, but also fear he may not have given full justice to his striking results. I would like to sketch what I remember."

The audience became transfixed, being unexpectedly treated to one of the "Oppie talks" for which he was famous. In a few flawless sentences one could print as they were spoken, he was able to summarize every seminar he attended and made the speaker see—often for the first time fully—what had been accomplished and should have been shared with the audience.

As he sat down, the mathematician John von Neumann, father of the computer, stood up. "I invited Dr. Mandelbrot to spend the year here, and we have had very interesting conversations. If he allows me, I would like to sketch some points that Oppie did not mention." The transfixed audience was then treated to a "Johnny talk"—equally compelling, and delivered with a strong Hungarian accent. The meeting went from abysmally low to unforgettably high and concluded in triumph.

Am I describing a nightmare? No, but I wish I was. Having left MIT, I was spending the year 1953–54 at IAS as the last postdoctoral fellow that von Neumann sponsored. That lecture came about one day during a chat with Oppie on the commuter train.

John von Neumann

Many pure mathematicians I knew well—like Szolem or Paul Lévy—were not attuned to other fields. John von Neumann (1903–57) was a man of many trades—all sought after—and a known master of each. He continually stunned the mathematical sciences by zeroing in on problems acknowledged as the most challenging of the day, and with his speed, intellectual flexibility, and unsurpassed power, he arrived at solutions that encountered instant acclaim. He did not seem to consciously search for any single holy grail or Golden Fleece of the mind beyond his readiness to tackle many diverse investigations. From the most abstract foundations of the purest mathematics to strategic advice to U.S. presidents, moved by insatiable curiosity and aided by personal wealth, von Neumann let his fancy run free. As soon as he heard a field had become hot, he made himself an expert with a competitive edge and identified several key issues he could solve.

Von Neumann had a "normal" childhood. So did other Hungarians in that celebrated age. Cohorts Eugene Wigner (1902–95) and Edward Teller (1908–2003) also achieved high fame in the United States and a substantial—though less flamboyant—level of versatility in combining abstract skills with interest in applications. The glittering culture to which they all belonged vanished after the Hapsburg double monarchy collapsed in 1918 and Hungary lost half of its historical lands. Thus, their development was thoroughly disturbed by an external element. Von Neumann started in the 1920s with a fundamental Ph.D. thesis in logic, specifically, abstract set theory. Next he did two great pieces of work, which I knew well. He first formalized the foundations of quantum mechanics. Before his work, two approaches had been in competition. In appearance, they were very different, but he showed them to yield identical results. Later, he "invented" the theory of games, which he meant to provide a foundation for economics. Then—still very young—he proceeded to other works that made him famous as a pure mathematician.

By the time I met him, he had long left pure for applied mathematics. Fascinated with weather predictions, he had become convinced that theoretical meteorology would remain primitive until the underlying mathematical equations could be solved numerically. To solve them, he had reinvented himself as an entrepreneur in an entirely untried form of engineering, and closely supervised a team building one of the first electronic computers—from scratch.

Inherited wealth saved him from ever working in a garret (figuratively or otherwise) or fleeing for his life—though for the hundred days of the Bolshevik dictatorship of Béla Kun, his family prudently left his native Budapest. Concluding that he would never achieve a professorship in Europe, von Neumann moved to professorships in Princeton, long before Hitler's rise to power, first at the university, then at the IAS—the most desirable of all academic institutions. He also became a highly paid consultant.

In truth, I disdained the nature of his interests and the fact that, while multiple unrelated interests made us fellow throwbacks, he was the precise opposite of a self-motivated solo scientist. As I already mentioned, the "hot" specialties that attracted him were overflowing with skilled competitors, and he was a formidable visiting expert who did not threaten his hosts. At that stage in life, I did not seek competition, but craved variety. He filled me with admiration, awe, and the desire to emulate the sheer vastness of his pursuits. I was also hoping to gain hints about how he managed.

Von Neumann's diverse interests continue to thrive separately from one another. The nearest thing to a proper centennial celebration was held in his native Hungary. Von Neumann was lucky that his country of birth—a very small nation—finds continuing solace in the greatness of its sons who went away and achieved fame abroad, therefore absolving (or enjoying) their idiosyncrasies.

Warren Weaver Saves the Day, More Than Once

Naturally, I had sent von Neumann a copy of my Ph.D. thesis. He sent word back that I should come see him—any day, even on a Saturday morning. While at MIT, I took a few days off to pay him a visit.

Very well dressed compared to some other academics, he looked like a banker. We talked and he asked if I could visit for a year. I said that I would love to, but when? It was late May, and I assumed that everything was settled for the next academic year. He responded that the Rockefeller Foundation in New York could easily solve this problem. On the coming Monday, I should see one of the great movers and shakers of scientific policy during World War II, Warren Weaver. Von Neumann would leave a message on my behalf, and everything would be settled in no time.

On the forty-ninth floor of 49 West Forty-ninth Street, in New York City, the receptionist waved me toward Weaver's secretary, who waved me into his office. On my way out after a brief but very nice chat, I asked for an application. None was necessary, I was told. Everything had been arranged. No major turn in my entire life proceeded more smoothly.

Over the years, I saw Weaver every so often. He always bubbled with new projects. At one time, he was committed to helping launch mathematical biology, wanted me to take a lead, and offered substantial funding. But I felt—correctly, it seems—that the field was not yet ripe enough for me to abandon my other activities.

My last encounter with Weaver, in 1968, was very different from the first, but equally unforgettable. By then, I was working at IBM. I had just started reporting to an individual who made my life difficult. The IBM policy at the time was never to fire anybody, but this new supervisor could easily hound me out by assigning some project that I would simply hate.

Fearing that the end was coming, I went to see Weaver, who was then at the Sloan Foundation. He revealed that years earlier "Johnny" (then dying of cancer) had asked him to keep an eye on me—he saw that my chosen path was dangerous and I might need help. So Weaver offered me a two-year fellowship as a visiting professor at a university of my choice. He also suggested that this money could find other uses, so I should first try to settle my differences at IBM.

Observing my surprise at these revelations, Weaver disclosed other significant facts. Von Neumann had long been unhappy at the Institute. Many mathematicians resented him for leaving "real" mathematics for computers. Mathematicians and physicists detested his well-known hawkish military views. In a way, as long as he was a pure scientist among pure scientists, he could impress the "natives." When he moved on to engineering and politics, the tolerance ended. As I found out, during the year I was at the IAS, he had accepted a position at UCLA—less prestigious than Princeton, but also presumed to be less stressful. He died too soon to find out.

I was so relieved at Weaver's offer that I did not question him further. How did my case come up between them? What other untold details of his story were lurking? Ignorance was bliss.

Back at work, the storm that I had feared soon dissipated, but I am grateful it led me to witness this extraordinary offer of help from beyond the grave. Von Neumann was not exactly a warm person, but (maverick to maverick?) he understood me.

A Commuter Ride with J. Robert Oppenheimer

One day, having boarded the train from Princeton to New York, I was quite pleased when J. Robert Oppenheimer sat down next to me. After scanning the newspaper, he turned to me. "Have you not just arrived from MIT? Please tell me about your work." That work was my Ph.D. thesis. Delighted, I proceeded to sketch it. He got my point instantly, confirming the observation by the physicist Hans Bethe that "Oppie" *could often understand an entire problem after he heard a single sentence,* and the observation by the physicist Robert Wilson that *in his presence, I became more intelligent, more vocal, more intense, more prescient, more poetic myself.*

I had hesitated to insist on the role of thermodynamics in the context of a social science—a topic other physicists tended to scorn. To the contrary, surprised and impressed, he told me, "Everybody tries to apply thermodynamics to social science problems but fails; you have actually achieved something."

He was especially thrilled to hear that my story of the Zipf-Mandelbrot law of word frequencies involved the notion of temperature of discourse. This fundamental exponent is usually greater than 1, but in certain special cases is smaller. In the theory of heat analogy, this meant that the temperature could be less than zero! A fact I thought had no counterpart in physics. Oppie interrupted in a very excited tone. "Indeed, it used to have no counterpart, but let me tell you about physicist Norman Ramsey at Harvard. His very recent work involves problems in which a negative temperature is not only unavoidable but very important."

Oppie ended by asking for my help. "I have been trying to organize evening lectures for 'the historians and the ladies' but find too few suitable speakers. Would you be the first?" I took a deep breath and agreed.

Ordeal by Fire: The Lecture and a Good Recovery

For days after Oppie's secretary fixed a date, I sweated to write a talk totally devoid of formulas and long words I might not enunciate clearly.

On the day of the lecture, I was in the room ahead of time and—to my horror—watched several Institute giants join the audience. Oppenheimer came in. "You need not come; you have heard everything I have to say!" "Not necessarily, and I want to be present." Then von Neumann came in. "You need not come; you have heard everything I have to say!" "Perhaps, but the discussion may be interesting. Besides, I am the chairman."

I trembled in fear throughout my lecture, watching famous people in the audience fall asleep and then snore. After forty-five minutes of agony, I called it a day.

Von Neumann stood up. "Any questions or comments?" Two friends commented and questioned me dutifully. As the gruesome experience was about to end, another man stood up. That's when

Otto Neugebauer proceeded with the blast reported in this chapter's first lines. Everyone was wide awake.

Through the night that followed, I was profoundly happy, but a question nagged me. I trusted that my worth and pitiful misery contributed to the obvious enjoyment Oppie and Johnny had both found in defending me. But could there be another reason? An answer came shortly after, when *The New York Times* publicized the gist of the celebrated trial in which von Neumann testified against Oppenheimer. Both wanted to go out on that night, and in very quiet Princeton, mine was the only show.

The next day, I visited Neugebauer in his office. He was very apologetic. "Please do forgive my outburst." "To the contrary, I come to thank you. Without your outburst, my two skilled lieutenants would not have been motivated to stand up to defend my work." The ambience became very pleasant, and he gave a demonstration of his astonishing craft. His research dealt with tablets that used the same cuneiform alphabet for a mixture of two different languages that once coexisted in Mesopotamia: Akkadian, which was Semitic, and Sumerian, of unknown origin. Therefore, each tablet could have either of two meanings, and identifying the proper one was a very difficult task.

To be so well treated by Oppenheimer was a high compliment. An institution that had Oppie in residence was automatically the center of living theoretical physics, a highly active field at that time. Hence, the competition in physics was ferocious, and the level of the junior members extremely high. Computers were so new that they were not yet rated, and the staff on von Neumann's project was not even considered academic.

The Name-droppers' Nirvana

I had visited numerous palaces, museums, and state monuments, but IAS was the first place I lived and worked where I was surrounded by elegance and gentility. Carva was a barracks, RLE's Building 20 prided itself on its decrepitude, and MIT's corridors suffered traffic jams between classes. By contrast, IAS seemed an oasis of motionless meditation; it even boasted of noiseless light switches, which were new to me.

One exception occurred every weekday at teatime. Practically everybody attended, except for the likes of Einstein and von Neumann. Clearly one era was passing and a new one was coming in, so between the great men and us low-ranked beginners, there was hardly any "middle." Most careers—including mine, for a long time—were doomed never to rise higher than this year at IAS. Thus, what should have been a marvelous experience was in many ways no fun at all.

Throughout the short IAS term, I had great fun, but I also worked diligently on many topics and obtained wide-ranging results. I presented everything I had done at a meeting organized by the Brooklyn Polytechnic Institute. When it came time for publication, logic and concern with a future career should have suggested "retailing" that work through several separate papers. Instead, I wrote a single long and involved "memory dump." I doubt that anybody ever heard of that meeting's utterly obscure *Proceedings*. For example, how long did it take for "normal" research to duplicate my findings? It was years before a formula I self-effacingly called Szilard's inequality really came out as the McMillan inequality of coding theory. Other formulas took decades to be duplicated. To my delight, a long and tedious calculation carried out in that paper proved its mettle by starring in a far more widely interesting context . . . in 1995.

* * *

I benefited—for life—from meeting IAS graduate student Henry P. McKean, who went on to a brilliant career. His thesis topic was pure mathematical esoterica. Puzzled by some complications and difficulties, I requested and received very useful coaching. The lesson was filed in my memory, and something called the Hausdorff-Besicovitch dimension of the values of a Lévy stable processes became essential to fractal geometry—which led to that dimension becoming well known.

17

Paris, 1954–55

Von Neumann was leaving Princeton for Washington, so I could not remain for the usual second year of a postdoc. Nearing thirty, I felt ready for a regular job. In the United States, a quick check yielded nothing I liked. France had no teaching job either, but excellent insurance.

Supported by the National Center for Scientific Research

Indeed, a research position granted by the CNRS had prudently been maintained. When I came back from Caltech, being drafted put me on unpaid leave. Later, Philips passed unnoticed, and that leave continued automatically when I was at MIT and Princeton.

I met the big CNRS boss in person and heard that not only a paying job was waiting, but also a promotion to *maître de recherches,* the third of four ranks from the bottom up. The CNRS was famously bureaucratic, and my official letter was anything but welcoming, mostly listing all kinds of prohibited activities. To the contrary, other recent Ph.D.'s received all-too-final letters of termination. Incidentally, later beneficiaries of CNRS largesse lived in a less generous system that is now further threatened. Tenure was immediate and ironclad, but promotion was glacial at best, and some remained at the lowest rank until retirement.

My rank of junior research professor was expected to last until a teaching position became open. To increase my chances, I volunteered to teach pro bono publico. Searching for a format one could not confuse with that of a "real" course, I settled on a misnamed *groupe de recherche* that resided in my briefcase and consisted of lectures on information theory that I gave and published. In research, I kept running around.

France happened to be abuzz with great political theater, thanks to its prime minister, Pierre Mendès-France (1907–82). Later, his mathematician son, Michel, told me about the origin of this name—a Portuguese ancestor named Mendes married a young lady named Francia. Fleeing the Inquisition, they moved to Bordeaux, where "Francia" became "France." His nickname, PMF, was a straight copy of Franklin Delano Roosevelt's FDR. Mendès had been an unusually young subcabinet minister before the war, then a wartime pilot, and next a minister of de Gaulle in London. Among the many French prime ministers between de Gaulle in 1945 and de Gaulle in 1958, he was rated best and remains most fondly remembered. But he was an incorrigible maverick and could never give full measure of his talents. Indefatigable, he was ridiculed by opponents for keeping a glass and a bottle of milk on his desk. In France? Yes, his constituency in Normandy produced milk, not wine. Besides, he was effective in fighting cheap alcohol.

Normal conditions would never have allowed him to become premier. But—shortly after a story told earlier, of the Carva contingent marching to honor the Vietnamese leader Ho Chi Min—the situation had become quite abnormal. Unthinking French governments had rushed into a war in Vietnam. It was doing very badly and encountering every problem Mendès had predicted. Ultimately, the fathers of the war—his unforgiving enemies—asked him to clean up their mess. This was the path followed by the old Ottoman Empire: whenever it had to abandon some territory, it just so happened that its foreign minister was not a Turk but a Greek. PMF delivered—helped by smoke and mirrors—then rushed to give up French protectorates in Tunisia and Morocco. At that point he was overthrown, and the political situation resumed the course that soon returned de Gaulle to power.

Close to home, Mendès took every opportunity to promote the sciences and bemoan their weakened state in France. He was widely heard, which may be why—before Britain or Germany—a wild and uncontrolled enrollment surge hit the French universities and the academic market flipped from puny to wide open. In a short time, this would have a major impact on me.

Paul Lévy

Getting to know Paul Lévy was one of my few academic accomplishments in 1954–55. He never had a formal disciple, I never had a formal teacher, and I never thought of becoming his clone or shadow. Yet much of probability theory has long consisted of filling logical gaps in his works, and in a real, though indirect, fashion, he was the teacher of several members of his family, and also mine.

He documented his life, thoughts, and opinions at length in a book well worth reading because of his lack of any attempt to appear better or worse than he was. The best passages are splendid. In particular, he describes in touching terms both his fear of being *a mere survivor of the last century,* and his feeling of being a mathematician *unlike all the others.* This feeling was widely shared. I recall John von Neumann saying in 1954, "I think I understand how every other mathematician operates, but Lévy is like a visitor from a strange planet. His own private methods of arriving at the truth leave me ill at ease."

When Lévy died in 1971, I lobbied for a memorial at Polytechnique, but very few people came. However, the centennial in 1986 was a different story. By then, Lévy's mistakes and idiosyncrasies were forgotten and forgiven, and a large meeting was organized by pure mathematicians. (A Polytechnique building came to be called Lévy.) Late in the process, I was invited, discreetly informed of strong opposition to my participation, and advised to avoid the shrillest opponents. Sadly, I wondered whether Lévy himself would have been invited and—if so—would have felt comfortable. I did not.

Lévy was the least flashy person on earth, so how to explain the profound influence his work and manner had on me and on many other scientists? Herein lies a familiar and always surprising story concerning the very nature of probability theory.

One half of the story is part of the mystery the great mathematical physicist Eugene Wigner called the *unreasonable effectiveness of mathematics in the natural sciences*. A symmetric mystery should never be forgotten: the unreasonable effectiveness of the sciences in mathematics. Together these mysteries acknowledge that human thinking is unified within itself (and even with feeling), not in a trendy New Age fashion but very fundamentally.

Georg Cantor claimed that *the essence of mathematics lies in its freedom*. But mathematicians do not pick problems from thin air for the pleasure of solving them. To the contrary, a mark of greatness resides in the ability to identify the most interesting problem in the framework of what is already known. And the highest level of the label "interesting" is invariably accompanied by a restrictive label, such as "in mathematics" or "in physics." My admiration for Lévy's "mathematical taste" increases each time his mark is revealed on yet another tool I need when tackling a problem in science that he could not conceivably have had in mind.

What a contrast with the period around 1960! Then Lévy stability was viewed as a specialized and uninteresting concept. It received at most a page in textbooks, with the exception of one by Boris Gnedenko and Andrei Kolmogorov. The English translation expresses the hope that Lévy stable limits *will also receive diverse applications in time . . . in, say, the field of statistical physics*. But no actual application was either described or referenced—until my work.

Lévy's minicourses—I attended several—have marked my whole life. Not a charismatic lecturer, he looked frail and withdrawn. The auditors were few, and I recall (wrongly, I hope) having often been alone. I also watched Lévy closely at the weekly seminar on probability. One speaker began by describing a problem on the blackboard, then faced Lévy squarely and invited him to guess the answer. The guess was correct. But how reliably could Lévy proceed beyond guesses? A book by Kiyosi Ito and Henry P. McKean is pointedly dedicated to Lévy, *whose work has been our spur and admiration*. It includes this comment: *The difficult point of this proof is the jump between [two equations on that page]; although the meaning is clear, the complete justification escapes us.*

Andrei Kolmogorov

A giant of my teachers' generation, the polymath Andrei Nikolaevich Kolmogorov (1903–87) lived in Moscow. Had it been possible, he would have joined Wiener and von Neumann in influencing my intellectual growth directly, but the Iron Curtain was then an insurmountable barrier. Like Lévy, he was celebrated for work in pure mathematics. He also thought about many aspects of the real world, including the structure of Russian poetry. In the 1930s, he obtained results in genetics that became textbook material. But he antagonized the notorious Trofim Lysenko, a quack favored by Stalin who destroyed genetics in Russia, and fell into disfavor. He reemerged with a pathbreaking paper on turbulence that we studied at Caltech and that was to have a direct influence on my research.

To everybody's delight and surprise, political maneuvers allowed Kolmogorov to spend the spring of 1958 in Paris. At a packed and unforgettable colloquium, he outlined the results of the work of two of his students, who both went on to great fame: Vladimir Arnold and Yakov Sinai. Arnold's results added tangibly to a major issue I would contribute to over the years—the distinction between objects of different dimensions. The first square-filling curve, demonstrated in 1890 by Giuseppe Peano (1858–1932), showed that a continuous motion can visit every point in a square. "Intuition" claimed that one- and two-dimensional objects did not mix; hence, in 1890, a plane-filling curve was called monstrous. That scandal lasted until fractal geometry transformed that monster into an intuitive tool. Arnold's results revealed to us in 1958 by Kolmogorov showed that every continuous function of a point in the plane can be expressed by combining ten functions of a point on the line. There must be a catch somewhere! Yes! Those ten functions have to be special fractals—long before the word was coined.

Kolmogorov had coauthored a textbook featuring an obscure mathematical object universally regarded as a mere toy, one that I later called Lévy stable probability distributions. The only real-world application in the literature was quite isolated and did not lend itself to development. But I was going to change that toy into an essential tool in economics. However, I was concerned about a sentence in that textbook. So I went to see Kolmogorov. My results obviously surprised him, and he praised them warmly. Then I asked for references to the precursors claimed in that textbook. He changed the subject. My suspicion that those references never existed was confirmed.

Wiener's nonobvious motivations are described in his memoir. John von Neumann seemed to seek the hottest topics of the day. What about Kolmogorov's motivations? A 1962 talk he gave in Marseille on turbulence was raw, and he never followed it with a piece true to his high standards. So when Russians close to Kolmogorov came west, I inquired about the motivations of that paper on turbulence. Again, I received no answer.

I still think the issue is important from the viewpoint of the unity of mathematics and continue to hope that a well-informed and bolder soul will educate us. I would also welcome the story of how an orphan from a small ethnic enclave of Russia rose to such a level of admiration and respect.

18

Wooing and Marrying Aliette, 1955

CHRONOLOGICALLY AFTER MOTHER, the most important woman in my life has been Aliette. Of course. We met in October 1950, shortly after my release from the air force. We did not rush to join our lives, marrying after five years of acquaintance. So, no matter how you count, our golden anniversary has passed. Being ill defined is a feature common to all important concepts.

Following a custom on its way out, my parents had not married until Father had settled down as a reasonable provider. My own slowness in settling down was acutely on my mind, but as I watched friends and my kid brother marry, it was obvious that I should rather wait.

My activities in 1954–55 left me plenty of time to woo Aliette, the second cousin of my Caltech classmate Leon Trilling. He had a traveling fellowship for 1950–51, decided to use it in Paris, and asked me to find an apartment he could rent. This I did, and was invited to the housewarming, where I met members of his family who had survived the war in France. This included his first cousin and her daughter, Aliette Kagan. Aliette was eighteen, a recent high school graduate registered as a student of law. She eventually changed to biology. I also met her brothers. Years later, the older received the Wolf Prize for chemistry.

The Trilling family had been prominent in Białystok, a Polish city halfway from Warsaw to Wilno. Family legend claims that Czar Peter the Great (1672–1725) brought their ancestor to Russia when he invited diverse experts from Holland to Westernize his empire. Therefore, they were merchants of the First Guild and could live or travel where they wanted in the empire. Typical of Russian upper classes, their first language was French. By 1939, Aliette's branch had moved to France, where everyone adjusted well.

I had confessed to my future wife that I had a very demanding mistress I did not intend to abandon. She did not mind. That mistress was—and is—science. Throughout, my wife has been extraordinarily supportive. Without her willingness to let me gamble my life—and hers and our children's—the odd career I undertook would have been unthinkable.

When we were getting to know each other, our evenings out were almost entirely musical. Soon after we married, we attended the most unforgettable opera performance of our lives: Mozart's *Don Giovanni* in the Monte Carlo Opera House, a miniature of the Paris Opera designed by the same architect, Charles Garnier. The marriage of Prince Rainier of Monaco to a famous actress called for the ne plus ultra of gala performances, which filled the opera's first rows with recognizable faces and fanciful evening dresses. We happened to be driving by and—on a wild whim—stopped to inquire. We were told that if we promised to keep quiet and invisible, we could be seated in the last row. The seats were cheap enough, and that last row was the thirteenth—close enough. The greatest singers of the day and a small but top-notch orchestra—nirvana!

Honeymoon at the Divine La Boverie

Aliette and I could not forget Geneva. Visiting in search of a place to rent, I had noticed a beautifully written newspaper ad, a poem in praise of a house in what seemed to be a distant suburb. I telephoned to ask for directions to Satigny and was instructed to just wait in front of the railroad station and watch for a car one could not miss, an Alfa Romeo 1800. In no time, a man with a small boy arrived. He introduced himself as Marbot, and the boy as his grandson, and took me to his estate.

The divine La Boverie consisted of a huge park with a manor house subdivided into apartments, a "farm," and a plain little apartment for rent above the garage. The estate dated to the eighteenth century— the name suggests that at one time oxen were kept there—and I was

Here are pictures of us in the Alps in April 1955, shortly after becoming engaged and on November 5, 1955, at our wedding celebration.

soon told that, to make the design last for eternity, the views were framed by sequoia trees, then a recent novelty in Europe. The same architect also featured sequoias in the big Parc de la Grange in Geneva.

A vineyard covered the sunny hill across the road. Together with the wheat fields, it had belonged to the bishops of Geneva, later overthrown by Calvin. Both lands had been said to be under the bishops' *mandement*, an ecclesiastic term with the same root as "command." Therefore, to this day, these lands continue to be called *pays du Mandement*. The high-sounding wine brand Perle du Mandement was plonk mostly used to dilute the far better Swiss wines from cantons farther east.

A deed was signed and—to show my fiancée that she was going to live in a grand place—Marbot gave me an aerial photo taken by an enterprising pilot. The view was as divine from the air as from the ground. Compared to the small apartment we would have had to settle for in Paris or downtown Geneva, this was a winner. Not like my parents' apartment, with windows on an alley, or a house like in Piranesi's *Carceri*.

We spent a two-year honeymoon there, and it is where Aliette brought Laurent, our older son, home from the maternity. It evokes a flood of memories. An immense lawn had to be rented to a farmer to plant wheat. That field hid a proper cherry orchard with fruit ranging

from small and tart to big and plummy—making dessert into a glorious many-course banquet.

Our apartment had a well-framed open view across the Rhône, with the Mont Salève to the south in full glory; in rare clear weather, we even saw the Aiguille du Midi in the high Alps near Mont Blanc. From the manor house, the view to the west went straight through Bellegarde, where the Rhône cuts across the Jura Mountains.

La Boverie, in the cool and discreet style of Calvinist Geneva, spoiled us for life. Each new shelter we considered had—in its own way—to match our first. Our future house in New York instantly attracted us with its oak tree reminiscent of "our oak" in Geneva. And later, when our younger son, Didier, insisted on a dog, we found for him a brindled boxer—dark brown with a black mouth. Magnificent thoroughbreds demand proper names, and we chose Bruno Boccanegra de la Boverie.

Our first car was basic: a Citroën 2CV, the fabled Deux Chevaux, which our friend Mark Kac called the Platonic essence of a car. Of the innumerable cars I owned, that alone deserves mention. Rolling down the canvas roof made it into a roadster; it could never be called dirty because at that time it only came in one color: dried mud.

"2CV" would seem to stand for the puny power of a two-horse team (CV meaning *cheval-vapeur*, or horsepower). But it doesn't. As soon as cars appeared, the government imposed an excise tax meant to increase with the engine's power. But horsepower was subject to argument and fraud. Hence, the law froze the relation between engine displacement and power that had prevailed when a "normal" engine of about 400 cc could produce 2 CV. The regular model had 375 cc, and our "luxury" model had 425 cc—less than most motorcycles.

André Citroën was a highly educated, very sophisticated, and daring innovator, both in design and advertising. He tamed front-wheel drive for mass production, and his brilliant engineers rethought every part from scratch so that even some key parts could be duplicated, if needed, in a home garage. The result was quirky in the extreme. One day, our car stopped dead in the Alps. I opened the hood and was mystified by a small part covered with grime. Once cleaned, it turned out to be a lever, which I played with just in case—and identified as an auxiliary fuel pump! A few pumping stops got us back home, and I rushed to the garage, which was housed in an old smithy. The proprietor told me that one part of the fuel pump was underdesigned, but there was no need to order and wait for a spare. On the spot, he machined a replacement from a chunk of steel scrap picked from a big barrel.

185

19

In Geneva with Jean Piaget, Mark Kac, and Willy Feller, 1955–57

MY POSTDOCS AT MIT AND PRINCETON had been carefully laid out, and the research position that followed in Paris was planned to support me while I was waiting for an academic opening. Unlike those periods of the grand tour, the 1955–57 stage of my "career"—in Geneva, Switzerland—was completely unplanned.

Jean Piaget

In 1955, the Institute of Statistics of the University of Paris was squeezed into a few rooms in the Institut Henri Poincaré, part of a small campus on what is now the rue Pierre et Marie Curie. One day when I had to stop by on some administrative business, a spry but elderly looking gentleman breezed in and asked the secretary where and when he could find me. Having found me quite easily, he introduced himself as Jean Piaget (1896–1980). He was pleased to hear that I was aware of his fame in trying to bring rationality to child psychology. He had long been a professor in Geneva, and at this point he also taught a day each week in Paris, commuting by overnight sleeper train.

We sat down to chat, and he described the notion that the nature of knowledge could be inferred from the way knowledge was acquired in early childhood—something he had studied all his life and called genetic epistemology. He had a Rockefeller Foundation grant to establish an interdisciplinary center that he was sure would move at lightning speed—if he had help from a suitable mathematician in residence. He was looking for someone whose work showed open-mindedness, was impressed by my work in linguistics, and wanted me to be that mathematician. I was thunderstruck. Only a few days before, Aliette and I had decided to marry, and we were both keen to live neither too close nor too far from our mothers in Paris but did not know how we would be able to do that. Piaget's sudden offer provided a timely, surprising, and most elegant solution. Our "negotiation" was brief. Yes, I could be an assistant professor at the university, but most important for him, I should be very active in a weekly get-together of all the participants and a broad symposium at the end of the year, with immediate publication of our results.

A truly ambitious program but—at my stage in life—a godsend! Geneva was close enough to Paris for me to keep an eye on openings in France. Piaget seemed like an interesting person, and working with social scientists looked challenging—and might help me land a job. I accepted, and Piaget attended my wedding party.

From a home up in the hills, Piaget biked to office and classroom, downhill or uphill, sunshine or rain. Therefore, his face was weathered, but he was young in spirit and years. His Ph.D.—earned when he was twenty—concerned mountain snails and familiarized him with the scientific practices of zoology. He promptly changed fields and set out on a lifelong effort to extend proper scientific principles to human behavior.

His first books on children's intelligence were based on his observations of his own babies and written when he was in his early twenties. Not resting on his laurels, he was always at work on papers, reports, or a book. Early in the school year, he asked me to look at his current book and handed me a chapter. I found it interesting but asked him to explain a few lines in greater detail. Piaget apologized and obliged: in no time obscure lines became obscure whole pages. It soon became clear that—until that moment—he had never heard the words "I do not quite understand. Please explain."

Before founding his Center for Genetic Epistemology, Piaget had achieved international fame while leading a completely sheltered and very austere life. He had mostly interacted with either students of education—awed and in a hurry to be certified—or confirmed schoolteachers who would never dare contradict him.

While Piaget could be vague or wrong, he was not a phony, and I always perceived in him an element of genius. Due to extreme isolation before the 1950s, his scientific talent had never been honed by competition. His ambition was boundless, with no inkling of the deep truth I had learned from John von Neumann: that a scientist shows mettle by identifying problems that are neither too easy nor too difficult. Science is best at giving credit for thinking big, but not too big. I worked hard, but sparked no miracle.

I admired Piaget's ambition to become the Kepler of psychology—but not his expectation that, with my help, a year or two would suffice. His center continued for years, and reportedly my successors worked out better than I had.

Mark Kac

By extraordinary good chance, my years in Geneva had a plus: close acquaintance with two other visitors who happened to be the most active probabilists on the west side of the Iron Curtain. In 1955–56, Mark Kac (1914–84), a Cornell mathematician, was in residence. He had no built-in group of associates, and there was no obstacle to our becoming close. The 1956–57 visitors were Willy Feller and Joseph Doob (1910–2004), whom I knew less well.

Mark Kac was quick-witted—always the life of the party. His storytelling skills were well above the typical mathematician's, and he was tireless in advocating greater harmony between mathematics and science.

His personal style, likes, and dislikes did not in the least match his dry-as-bones articles. He had been deeply influenced by his teacher and spiritual father, Hugo Steinhaus (1887–1972), a mathematician who had trained in Vienna around 1900, at a time when it was a major intellectual center. His ideal was not too far from what Hadamard had accomplished, Szolem had spurned, and I was hoping to achieve: a harmonious alloy of mathematics and science.

But life was dangerous in Poland, and the first order of the day was to find a way out. A fellowship to Cornell eluded him in 1937. He was bitterly disappointed, but—as he later gleefully told anybody who cared to listen—fate had been on his side. If successful, he would have returned to Poland in 1938—and likely perished in the war.

The appointment did go through for 1938, but the letter from Cornell was adamant: under no circumstances could it be renewed. He had hit a wall and was ready for anything. He watched the content and style of the conventional mathematics that were in favor at Cornell. Then, contradicting the openness inherited from Steinhaus, he simply morphed—for the duration—into a follower of fashion. Conversions under duress were very common during the Depression—as I knew from the case of Father.

Our turbulent childhoods made us react very differently. He gained high respect for order and fear of anarchy. One day when we were chatting after a lecture, another attendee came up and expressed delight at seeing two mavericks together. Smiling as usual, Kac responded: "Benoit is a true maverick, but I am not one in the least. I am a staunch conservative who tries to act intelligently."

Years later, he influenced my life by firmly telling me that, instead of more papers that looked unrelated, I must write a book. So I did; my first was in 1975, in French. He reviewed the later English version of 1977 favorably. But in private, he expressed fears that I would open the gates to a flood of nonsense—fears I shared but had to face.

William Feller

I first met the mathematician William Feller (1906–70) in Paris; I next saw him in Princeton in 1953–54, then in Geneva in 1956–57, and later—repeatedly—when he consulted at IBM. He deserves a few words here not because he became a role model for me—he definitely did not—but because he was expected to become one.

Let me begin with a quote from a tribute to Feller by Joseph Doob:

Those who knew him personally remember Feller best for his gusto, the pleasure with which he met life, and the excitement with which he drew on his endless fund of anecdotes about life and its absurdities,

*particularly the absurdities involving mathematics and mathemati-
cians. To listen to him lecture was a unique experience, for no one else
could lecture with such intense excitement.*

Feller had been a prodigy at the University of Göttingen and
earned his Ph.D. at age twenty. Feller's paternal grandfather was Jewish, so he had to leave Germany. The Depression brought him to
Stockholm, under Harald Cramér (1893–1985). Cramér loved pure
mathematics but owed his funding to strict Swedish regulation of the
insurance industry and had to satisfy his benefactors. So did Feller.

When in Sweden, and later as a colleague of Mark Kac, Feller
became an effective teacher of probability theory, and his marvelous
textbook was beloved by many scientists who trusted that mathematics was of genuine use in the sciences. But astonishingly, Feller went
out of his way to pooh-pooh this trust. In a published interview, he
described as fraudulent the idea that the famed bell curve of mathematical errors ever represents anything real. He even denied it had
any role in what is called thermal noise, where it is a pillar of excellent
theory and unquestioned practice.

Probability saved his career and made him rich, but it was never a
true love. It was a stopgap until he could return to purer mathematics
by leaving Cornell for Princeton—then in a golden age.

My 1962 pioneering work on the price of cotton and other com-
modities had been my first Keplerian jackpot. Soon it was pushed
away in horror. Its detractors included Feller. When I submitted an
early paper on prices, IBM asked him to comment. He flattered me by
praising a technical angle, but he proclaimed that what I did had nothing to do with the real world. This made my IBM manager very
unhappy, and to save myself, I had to exhibit Feller's infamous article
about the bell curve and thermal noise.

My work on cotton prices was followed by work on the ebb and
flow of the Nile. The brilliant Harold Edwin Hurst (1880–1978) had
discovered a relationship that everyone characterized as a deep riddle.
Feller credited Hurst in a paper, but immediately proceeded to tackle
a related topic that he could handle, one that led to new mathematics,
yet was distinctly traditional, while Hurst's was not.

After the Hurst-Mandelbrot theory had solved the empirical riddle, I asked Feller to stop by my IBM office during one of his visits. To
be frank, I set him up. He began by restating his belief that Hurst's riddle could be resolved in a way his paper had suggested. I ventured that
he did not think this riddle had much bite. He conceded, with a smile.
Only at that point did I reveal my solution and its consequences—
both theoretical and practical. He got the point and became uncharacteristically subdued. Never did he accept me, but at least he ceased
to be an albatross.

20

An Underachieving and Restless Maverick Pulls Up Shallow Roots, 1957–58

THE SUMMER OF 1957 was scheduled to mark the end of the grand tour apprenticeship. My postdoctoral experiences in Cambridge, Princeton, and Geneva had been absolutely crucial to my personal and scientific development. Unfortunately, my various enterprises up to 1957 had not gone very far to further my aging but still vibrant Keplerian dream. The start of the academic year 1957–58 was supposed to be the beginning of "real" working life as a French academic in Lille and Paris.

Returning to Paris from Geneva in the fall of 1957, I could not help but think back to the fall of 1944, shortly after Paris was liberated. Then, despite my grossly curtailed preparation, I was on the way to shine at the tough entrance exams at the tiny École Normale Supérieure and the École Polytechnique. I was the academic star of the year, and Uncle Szolem—being impressed—was doing his best to recruit me for pure mathematics.

Marvelous Surprises!

The academic year 1957–58 represented a development I had completely given up on. I landed a teaching job with ironclad tenure at the University of Lille, plus a lovely moonlighting slot at Carva and other attractive prospects in Paris. When I received my Ph.D. in 1952, French universities had few openings, and I had not made the short list for an academic slot. But in 1956, when enrollments ballooned, teachers were suddenly in great demand.

Scarcity was such that my nominal adviser, Georges Darmois, remembered I was available. He telephoned to Geneva, asking me to come back and fill a vacancy. I was already committed for another year but for 1957 gladly agreed to become a soon-to-be-tenured junior professor of mathematics. I chose the University of Lille—only two hours north by train from my home in Paris. Also, ten years after graduating from the École Polytechnique, I was invited—practically begged!—to come back "home," as a junior professor of mathematical analysis, untenured and on short contract.

So the husband of Aliette, the father of baby Laurent, and the new owner of a very nice apartment in Paris close to the beautiful Parc Monsouris was now also a university professor. Drawing two half-time salaries from the National Treasury was a privileged but fairly common practice. In Lille, my teaching largely fit in two successive midweek days, with only one night at a hotel.

Teaching at Lille

Officially, the state felt obligated to provide housing to every civil servant, but all they offered me was a mean worker's cottage in a distant suburb of Lille. I took one look and decided to fend for myself. Anyhow, Aliette and I were ready to live again in Paris—the usual attractions being enhanced by two grandmothers waiting to provide for our baby son. Therefore, the most desirable provincial university became Lille.

We lived south of the Latin Quarter. Every conceivable way to the Gare du Nord crossed the old Les Halles area of midtown, but—compared to what it is today—traffic in Paris was a sweet dream and my trusty Citroën 2CV always found free street parking a short walk from the station.

In other words, I was joining the ranks of the part-time *turboprofs*, with little social life in Lille. The locals never invited us and criticized us for being absentees and less available to the students.

Substantial welfare-state perks were paid only once, and one salary was withheld to repay the zero-interest loan for my apartment. Therefore, on top of job security, my financial situation in France was satisfactory. In addition, Darmois was close to retirement. Musical chairs would open a junior position in Paris. Candidates were few, and—amazingly—my chances of being chosen and saved from a commute looked excellent.

Even this was not all! A marvelous and completely unexpected additional "escape route" soon opened thanks to a celebrated historian. Fernand Braudel (1902–85) was best known for *The Mediterranean,* a sweeping masterpiece written from memory when he was a war prisoner during World War II. I had read with fascination his description of the 1571 naval battle of Lepanto, where Spain had prevented Turkey from taking over the whole Mediterranean—and beyond. Braudel's group of historians, the Annales school, wielded considerable academic power and felt, at that point, that the wave of the future was quantitative history. They held an overly enthusiastic interpretation of the Zipf-Mandelbrot law of linguistics and of my effectiveness in Geneva with the psychologist Jean Piaget. So they invited me to set up an ambitious research group in Paris, west of the Luxembourg Gardens.

So by the fall of 1957, I was a beginning assistant professor at the University of Lille. I was not much noticed by the powerful French pure mathematics establishment, which I had spurned in 1945. A career as a disaffected civil servant with axes to grind would have been pleasant enough. But safety was not my goal, so that very thought made me shudder.

The Glitter Wears Off and My Plans Change

The marvelous surprise of the previous fall had worn off fast, and I ended the year 1957 in a very unsettled mood. I saw no compatibility between a university position in France and my still-burning wild ambition and dreams.

For the experienced survivor I was by 1958, the omens for an intellectually satisfying career in French or U.S. academia looked grim. Not to mention that teaching—even in a university—is a hard profession—one had better start practicing much earlier than I did. Moreover, in May, heavy political clouds burst with the return of Charles de Gaulle to power. No one could predict that he was to feed the French universities and later leave them to the self-destructive devices of the system that went up in flames in the events that shook the general's rule in the May 1968 riots.

I escaped with minimal agonizing. That year changed character, and a summer job sent me instead to IBM Research in the United States. My midlife crisis led me to forgo ironclad French tenure for an unknown position in the United States. Would this position last? In many ways, my timing was perfect, and my career bloomed beyond my wildest dreams. Had I not chosen a very risky path? Indeed. In fact, I allowed risk to increase enormously. Instead of joining any existing community of scientists, I went my own way and kept moving into topics that were not part of any field or establishment.

On July 20, 1958, unaware of what I was doing, I cast a die never to be retrieved. A summer job at IBM in New York ended my apprenticeship.

Part Three

My Life's Fruitful Third Stage

*The next several chapters describe highlights from the thirty-five years
(and a few days) when I was employed by IBM, and the partly
overlapping seventeen years when I was a professor at Yale.
I remember those years as a golden age—for IBM, for me, and
for the sciences. Let me go further, raise my voice, and add:
and also for the human spirit. What makes humans human
is the ability to speak—and to innovate.*

*I found fulfillment in seemingly disparate topics that did not
follow any usual pattern, were widely received as bizarre, and
often unwelcome. No institution I know of might have
replaced the IBM that I knew.*

21

At IBM Research Through Its
Golden Age in the Sciences, 1958–93

JUNE 20, 1958, was originally meant to begin a summer visit to IBM, located in Yorktown Heights, about an hour north of New York City. Aliette, baby Laurent, and I flew from Paris to New York. The three of us and our few suitcases easily fit into a taxicab. The difference in time zones made that day very long and tiring—yet almost routine and humdrum. But appearances were deceiving.

A Summer Job Becomes a Life's Work

Chance took me to IBM for that summer job in 1958. But only weeks after my arrival, I changed my mind and decided to stay. When I retired in 1993, my formal link to IBM had lasted for close to the thirty-five-year stretch of that company's concern with science. I stayed neither for the income nor for any subtle reason. I stayed by simple necessity—as I interpreted it after a staff meeting called by IBM's first director of research, Emanuel Piore (1908–2000).

Manny started the meeting with an observation: "I hear rumors of great unease among the troops. Many of you seem to wonder why we hired you, and you worry about the constant churning. When will the management stabilize, and when will you be told what we really want you to work on? In fact, there is no secret whatsoever."

As he continued, he confirmed my own interpretation of the situation.

"Most of you are fresh Ph.D.s and believe that the highest calling is to compete with your former adviser in adding footnotes to your thesis. But you'll soon find that on a daily basis, pure scientific research is a very difficult and in most cases unrewarding profession. At work you never have enough time to do what you want, and your wives complain that on Saturday mornings you go to the lab instead of taking the kids to the ball game. If you want to please your wife by being better paid and coming home in the evening without a briefcase, just tell us. The IBM development laboratories in upstate New York must grow, and their staff must be almost completely upgraded. Many prime jobs beg to be filled.

"But you may be hooked on pure scientific research. Fine. Wonderful. Research can offer you the choice between any number of exciting and well-rewarded tasks. Some of you may even dream of becoming great scientists. Marvelous! We can easily afford a few great scientists doing their own thing."

I still remember an extraordinary thrill and feeling of relief upon hearing those words. Given my gambling mood at that point in life, IBM's constant churning was a major attraction for me. The last thing I wanted was the order I knew in France. Right or wrong, I felt sufficiently stimulated by personal dreams. So I gambled on IBM and IBM gambled on me. We both succeeded and those successes were not unrelated.

Manny Piore's words convinced me—and, in turn, I convinced Aliette—that it would be best to linger at IBM for perhaps a year or so. Soon it became clear that I had set myself to remain there for an open-ended period.

★ ★ ★

I think back to Manny's words at that staff meeting: "Pure scientific research is a very difficult and in most cases unrewarding profession. . . . You never have enough time to do what you want . . . on Saturday mornings you go to the lab instead of taking the kids to the ball game." These words would come true in my case. The Hippocratic oath, taken by physicians, says, "First, do no harm." I strongly feel that it ought to apply to scientists as well. A father with a self-assigned and never-fulfilled mission is not a full-time father and can play havoc with his family. With Aliette in charge, I think I was able to abide by the oath. Let's leave it at that.

Settling in the United States

Now that we had decided to stay, it was time to find a place to live. Our two-year honeymoon near Geneva in the peerless La Boverie was in our minds when Aliette and I went house hunting. Lady Luck helped, and for four years we lived in an uncannily similar place—above the garage on an estate belonging to David Swope, where our second son, Didier, was born.

When we moved in, we missed the open view, south over the Rhône River, that graced our first home. But—once again—fate was on our side. The night before Thanksgiving brought a dreadful storm, and the next morning—lo and behold!—enormous trees had been uprooted and an admirable view had opened down the Hudson River all the way south to the Tappan Zee Bridge and, on clear days, even to Manhattan!

After the Swope estate, we bought a house in Chappaqua, chosen for its blandness—which delivered on its promise—and lived there for five years. It was a perfect setting for our young boys. In the photograph on the following page, I am flanked by Didier and Laurent in our Chappaqua living room.

For the next thirty-five years, our home was a place matching La Boverie and the Swope estate: a wonderful old pile in Scarsdale, five minutes from shopping yet so completely isolated from the neighborhood as to remind one of an old-fashioned Japanese house—on a far bigger scale. After an anonymous entry, the road immediately took a sharp turn into a lot so private it felt like a reassuring womb. Near the top of a rocky outcrop that probably never seemed worth farming, it became lost in a grove of very old oaks. The oak nearest the house may well have dated to the arrival of the white man to those shores.

Because the thorough complication of this house—which en-chanted me—scared off all "normal" bidders, we could afford it. It was built in stages between 1840 and 1940, so the height of the rooms on the first floor varied, and the second floor was filled with threaten-ing steps. Not one window was a rectangle when we moved in, as we found out whenever work had to be done. No contractor dared esti-mate the cost of any change. But chance led me to a soul mate, a man named Robert Robillard, a schoolteacher who knew how to use every tool—the kind of man whom I visualize in a horse-driven wagon on the way to conquer America. He needed additional income and was looking for intellectual challenge, and so for many years, the two of us fixed basic wear and tear in that old pile.

So the date June 20, 1958, has come to mark in my mind the mid-point of my life. That date has figured on thousands of forms and will not be forgotten. Next to the 1936 move from Warsaw to Paris, it wit-nessed the second major break in my life.

Unlike the first, this break never became complete, insofar as we still speak French at home. Besides, how could I possibly forget the country that helped me survive the war, accepted me, offered me its culture, and made me a free man? I never regretted the move to the

United States. However, increasingly—and unavoidably so, as time goes on and friends vanish—I feel in France like a visitor from far away and far back in time.

Emanuel Piore

Born in Lithuania, Manny Piore had received his Ph.D. when the Depression hit the United States. He survived the bad days, and when war broke out and scientists were suddenly needed, he went to Washington, D.C. He became a key man in the creation of a three-legged institutional system linking the National Science Foundation (NSF), the National Institutes of Health (NIH), and the Office of Naval Research (ONR) in a policy to support science well beyond topics of direct interest to the navy. Shortly afterward, he became director of research at IBM and built up the Thomas J. Watson Research Center.

In other words, this one remarkable man was responsible for several major science-funding institutions in the United States. Neither a great scientist nor an engineering innovator, he was a shrewd operator with a sense of noblesse oblige. The system he built was far from perfect but had an extraordinary asset: each leg used different criteria. And IBM Research chose its own way.

Since then, IBM and the ONR have abandoned pure research. The bulk of academic support comes from the NSF and the NIH. The process they use to select their beneficiaries applies the same criteria to everyone. This may explain why the NSF has always treated me abominably—almost invariably, the topic I proposed to investigate was perceived as incomprehensible, annoying, or worse.

Ralph E. Gomory

IBM's research staff was mostly very young in 1958. Shortly after my discreet arrival, a new hire thundered in—Ralph E. Gomory, five years my junior. For me, he was the most important IBMer in every way.

His Ph.D. from Princeton was also in a classic subfield of pure mathematics that he immediately abandoned as being too mature. He then solved a famous problem of applied mathematics by finding an algorithm that gives integer, not fractional, solutions to linear programming problems. Integer solutions are required in many actual applications, and solving this problem made Ralph extremely well known. IBM hired him and he achieved several other breakthroughs.

We met not long after his arrival, when my elder son joined a play group run by his wife. We became good friends, and I moved into his department. He was named director of mathematical sciences, then director of research, before he left for IBM corporate headquarters. After retiring from IBM, he was the president of the Sloan Foundation for many years.

Ralph's being my manager may well have been the longest-lasting direct or near-direct reporting relationship at IBM. Of course, as his schedule grew ever more demanding, our actual meetings became increasingly rare. But there is no question that, after Lady Luck, it is to Ralph that I owe the most: being made an IBM Fellow and, more important, being allowed the freedom to either wander off or dig in, without which the gamble I took would not have lasted nor borne the fruit it did.

Golden Ages—Mythical and Real

The golden age of IBM in the sciences coincided with the thirty-five years of my own belated golden period. It began roughly when I came to IBM and ended on a day when half of the staff was asked to leave and the other half was asked to get practical. Having come by chance in 1958, I stayed because nobody offered a better fit, and quickly thrived. The gamble for both parties was very successful. Within three years I had made two major discoveries—like hitting two jackpots in a row! Each one brought a visiting professorship in a field I knew almost nothing about: first in economics (later called finance) and next in the field of exotic noises.

My consulting role for "down-to-earth" colleagues was varied and mostly enjoyable. In many cases, I was solely a passing helper. But on one unforgettable occasion, a colleague and friend, Jay M. Berger, asked for help with some unglamorous but pesky noises on telephone lines. Our solution saved IBM from investing heavily in a development that was bound to fail. This episode also steered me to a scientific topic that I would have certainly missed.

Otherwise, consulting hardly interfered with my contributions to diverse existing sciences—or perhaps to a new science of roughness. IBM gave me the base that made everything else possible. It allowed essential visits to Harvard and Yale and other arrangements of brick-and-mortar. It also allowed me to move effortlessly between nonoverlapping fields of knowledge and organized science.

Does it follow that those years were a paradise on earth? Of course not! During IBM's years of glory as a scientific powerhouse, daily life saw many avoidable negatives. Of course! But remarkably, the overall balance was very positive. It may be that—to last—pure gold must be alloyed, and useful alloys produced in great haste may contain noxious elements. There was a constant element of pointless silliness, bad apples at different levels of the hierarchy, and the like. A paradise IBM's golden age never was. But I always felt that personal freedom, if not priceless, was very expensive. I paid the price and got something in return. Fair bargain.

How IBM Came to Invite Me

By my life's peculiar standards, the causal chain that led to IBM was short, with no unusual wrinkles. When I was at Princeton as a post-doc, I met Manfred Kochen (1928–84) at the Institute's cafeteria. He was a younger man with a background and ambitions similar to mine. He went on to join IBM when the brand-new research division was recruiting relentlessly and indulging in active public relations. Hearing I was going to spend the summer of 1957 at Cornell University in upstate New York, "Fred" brought me to a temporary site—ostensibly for a lecture but in reality for a combined interview and sales pitch.

His manager was the physicist Michael Watanabe (1910–93), whose Ph.D. chairman in Paris had been Louis de Broglie. The manager above that was Nathaniel Rochester (1919–2001), a career IBM engineer credited with a near replica of von Neumann's pioneering Princeton computer. Staff was needed for a machine translation project—very premature but well supported—and I was a rare warm body with good name recognition for my work in linguistics. I told them that a very nice job was waiting for me in Lille, and that my interests had shifted. Their answer was that they needed good people in every area, and they reluctantly changed their offer to regular summer visits, beginning in 1958.

Coming to Terms with Being an IBMer

In 1958, IBM was weighed down by an old and once carefully groomed reputation for extremely provincial and paternalistic human relations: company songs, compulsory white shirt and proper tie, and the like. Out of the blue, it set out to hire an entirely different technical workforce. As indicated by Piore's words early in this chapter, the research division often felt like a hiring office for the development sites in upstate New York. In earlier years, most IBM hires had come from small colleges or trade schools. A new flood from competitive schools of engineering created extraordinary change on a daily basis.

While IBM's permanent home was being constructed, the staff was moved to several temporary locations. The largest was in the village of Yorktown Heights. I was assigned to the much smaller Lamb Estate, Tudor-style buildings scattered around an incredibly beautiful site overlooking the Hudson River.

To someone who had lived through major events and read many history books, the atmosphere at IBM then recalled an aspect of France during the Revolution and empire. Since the ancien régime upper crust had mostly emigrated or holed up in provincial estates, very few promising individuals were available for promotion. Therefore, the selection rules had to change, and various old restrictions on inclusion were loosened.

For example, I think of Lazare Carnot, a French leader whose importance is often underestimated. An engineering officer trained before the Revolution, he was underused and feared early retirement. When the Revolution came, he was put in charge of building up the army and had to choose between timeservers who had not emigrated and men who in peacetime would have been passed over for their nonconformism, nonaristocratic or ethnic background, or other conspicuous "flaws." The Revolution succeeded because Carnot hired men such as the Corsican Napoléon Bonaparte.

For different reasons, IBM found itself in a similar situation. Competitors like MIT, Bell Labs, and General Electric, flush with money triggered by the Soviet Sputnik, were free to hire or import anybody with impeccable credentials. What made IBM Research a unique experiment—historically very significant if not always flawlessly planned? For one thing, relaxed hiring rules brought in many individuals for whom other institutions did not compete: "oddballs," "wild geese," scientists whose high-class record was marred by some fault or another or by disputes with faculty advisers.

I think of John Backus (1924–2007), who probably never had an adviser because he attended many schools—and none for very long. He contributed mightily to IBM's hegemony at one time. Using a computer early on was extraordinarily difficult and time-consuming. Every problem had to be broken down by hand into a multitude of very precise instructions that had been wired into the dumb machine. With a small group, no fuss, and ahead of schedule, John developed a "high-level" programming language dubbed FORTRAN (from "formula translator"), which was never a work of art or admired, but had an unarguable virtue: it existed. Compared to the earlier assembly languages, it was nirvana. IBM was lucky that he did not work for a competitor.

I think of John Cocke (1925–2002), who sounded and looked like a bad film's take on a rich senator from North Carolina. He arguably stood next to Seymour Cray in the tiny cohort of people who understood everything about computers; in particular, he originated something quite important called RISC computer architecture.

And of Gerd Binnig (very much alive), whose school record had been spotty but who impressed Alex Müller of the IBM branch in Zurich as having a mind capable of "lifting the heavy Swiss dough." He went on to invent the microscopes that can see atoms, to bring to IBM a flow of licensing fees, and to create nanoscience. He received for his efforts a Nobel in physics.

He was followed the very next year by Müller, who was awarded the Nobel for discovering high-temperature superconductivity— triggering a "physics Woodstock."

Many of these oddballs eventually settled down or left, but a remarkable several dozen remained. Their flawed, inadequate, or unconventional early résumés were forgotten, and for their contributions they harvested academy memberships, five Nobels, and other honors beyond counting.

Did this unplanned experiment prove anything? I can't be accused of envying those who do well at exams (nor of biting the hand that fed me) by noting that the IBM experiment confirmed my long-standing lack of respect for exam rankings.

Programming Before a Pervasive Concern with Security

For years, the research division owned no computer. A few hours a day, it could borrow one in Poughkeepsie, New York. The programs were punched on computer cards and transported—a several-hour trip—in a station wagon shuttle. A painfully awkward process actually worked. Your deck went in the morning to Poughkeepsie, and in the evening it returned—mostly with the message that some dreadful programming error had to be fixed. You sent it off again the next day, and so on.

One colleague spent an incredible amount of time on this process. To compute astronomical tables according to the Babylonians' model of the heavens, Bryant Tuckerman was sweating blood. I dared to wonder what the rush was. This calculation had waited for several millennia, and waiting for the faster computer soon to be installed locally would barely make a difference. Bryant stuck to his guns and produced an enormous document. Finally, a number of secretaries retyped the computer output in a form suitable for printing, and a huge book was published by the American Philosophical Society. I fear that very few copies were either sold or used. But these comments are anything but critical. The tenacity of my colleagues who first tamed the beast was an object of pure wonder.

Bryant also played a role in the droll story of how passwords came to Yorktown. Yes, there was a time when our computers required no password! My older son's math teacher was himself learning to program, and he introduced programming to his students. Having failed to make a certain program work, he asked my son to consult an expert—me! Begged to help, Bryant entered my account, wrote the program in no time, and printed a letter-size sheet for my son and his teacher.

A few months later, our computing center manager stopped me in the hall. "I am amazed. Of the substantial computer time available in the research division, you alone are using about half. I thought you were a very theoretical person." "I am equally amazed because weeks ago I gave up programming for myself." "So how come you are such a big user?" Monitoring revealed that I was billed for a mass of tiny programs run by high school students all over the surrounding Westchester County. At least one ingenious student or teacher had realized that simply typing my name in a box connected the user to the day's biggest computer—at no charge.

At that point, the computing center staff had to assign passwords. So I can boast—if that's the right term—of having been at the origin of the police intrusion that this change represented. Of course passwords must have originated in many places, and IBM Research's turn would have come very soon.

Computer Graphics Before IBM Was Involved

When my books—and then the fractal art—became known to seemingly everyone, my good eye was hardly ever praised. Instead, my merely being at IBM led to the perception that I was a lucky and passive beneficiary of an unfair competitive advantage.

In fact, I was not. Many other labs offered graphics off-the-shelf, but when I joined in 1958, IBM manufactured none, and to get outside equipment was harder than improvising. So on my way to achieving an unexpected status as a pioneer of computing—without ever touching the computer myself, but always giving instructions to programmers and assistants—I was forced to hustle ceaselessly.

Altogether, computer graphics took even longer than FORTRAN to become available. By the late 1960s, the most primitive helped me draw the first coastlines of artificial fractal islands. Our program simulated the whole relief. We could not visualize that "forgery," but visualizing the coastline was possible. We were working on a grid of sixty-four by sixty-four pixels, and the first step consisted of leaving blank all the pixels that, together with their immediate neighbors, were either above or below sea level. The untouched points defined an approximate coastline. The output device was an ordinary typewriter, and the idea was to print the points on the coastline by superposing the letters *M*, *W*, and *O* (or something like that). I copied this printed output and then blackened the inside of the island using a felt marker.

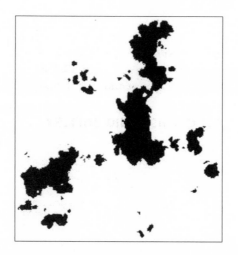

The process was heroic because the "software" needed to type the *M*, *W*, and *O* was not properly documented. Also, the buffer memory was tiny: having printed sixty-four bytes, the program stopped—until the word "return" was typed by hand. Desperate, I begged my assistant to type it again and again, as long as it took to get an output. When I went home at the end of the day, he stayed on. The next morning, the desired output was waiting for me.

A subsequent early device was manufactured by a company called Calcomp. It consisted of a sheet of paper rotating around a cylinder and a pen that could be lifted or lowered onto that paper but could only move along the cylinder's axis. A program combined the motions of the pen and the cylinder in an excruciatingly slow process, and the patterns it could draw were limited. We were pushing the machine well beyond its original specifications.

At long last, around 1970, graphic devices changed from mechanical to electronic. So a figure computed on the big mainframe could be delivered to a very shaky special purpose computer that allowed it to be examined on a laboratory cathode ray tube (CRT)—like an ordinary television screen. A special attachment made it possible to photograph the screen with a Polaroid camera. The first published pictures obtained in that way were the earliest fractal mountain reliefs.

A third graphics system, Lblgraph, came to life serendipitously in the early 1970s when IBM stopped an ill-starred foray into computer-based typesetting—and several colleagues and I changed it from black and white into sixty-four shades of gray.

To compensate for Yorktown's geographic isolation—and to help the world know that we existed—a steady flow of visitors were invited to lecture, entertain, and educate. No striking picture was ever shown, and I was dying to go to color graphics, but my immediate management turned me down: IBM was not in that business, and purchasing competitors' products was difficult.

One day in 1976, the grapevine reported that an outside supplier had installed our dream color graphics device for a colleague in the development group. On the spot, I called him and walked over for a chat. "Could we have access to this machine, and if so, when and how often?" "Of course. Gladly. But you must know that there is absolutely no software. I can pay a systems programmer, but to get one will take six months. Writing the software will take another six months. Come to see me a year from now." "Well, well . . . we actually are in a bit of a hurry. Could we perhaps have your lab's key code and stop over this weekend to get to know your gadget?" "Sure, why not." So before leaving on Friday, he gave the code to my very close colleague, Richard Voss, who went immediately to work, likely taking no time off for sleep. The software was ready on Monday morning. One year of waiting had melted into one weekend!

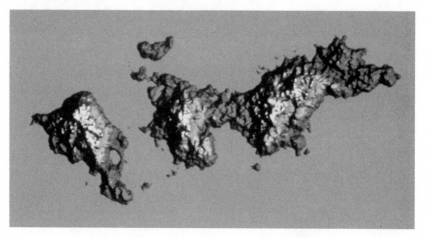

So, why study coastlines? Initially, I picked them because nobody had a permanent interest that would interfere with their acceptance, but also because my father was a map nut. From him, I learned to read maps before I could read and write. One of the most striking features of fractals is that they enable us to imitate nature. After the first general idea of coastlines, I thought of constructing random coastlines from a simple formula, and then random landscapes. Without computer graphics, it would have been a herculean task.

Within a university, that color graphics device would have belonged to the NSF through the project that paid for it. Therefore, its use would have been severely constrained. But at Yorktown, all the tools belonged to IBM and were assigned to projects as needed. That funding method had the advantage, within limits and with proper justification, of giving everybody access to equipment—if assigned to close enough friends.

* * *

In joining IBM in 1958, I resumed—on a far larger scale—a lifestyle I had once known in the past. The small Paris laboratory of Philips was replaced by the huge IBM Research, and an undemanding graduate school was replaced by academic nomadism: a sequence of visiting professorships in distinct and very different fields and of "traversals" of seemingly incongruous areas of research. They appeared at first sight to clash badly—but they really didn't clash at all. It soon emerged that I was working on the building blocks of my soon-to-be fractal geometry of nature.

22

At Harvard: Firebrand Newcomer to Finance Advances a Revolutionary Development, 1962–63

MY INVOLVEMENT WITH THE BEHAVIOR of financial prices—absolutely unplanned—became a constant of my scientific life. This revolutionary development went on to inspire many later works of mine, scattered around seemingly unrelated fields, and led me, in due time, to put forward a sharp distinction between two very different states of randomness: the "mild" and the "wild," and a third state I call "slow."

For several years—with IBM being focused on growth and the continual reorganization of its research division—I did little to be noticed. My first major piece of new work at IBM was embodied by a long publication, *Research Note NC87*, dated March 26, 1962. I was

acutely aware that my findings would have devastating consequences for the accepted standard theory of speculation. I was in a big rush to finish it—but I had low priority for secretarial assistance. I could not wait for it to be typed professionally, so I proceeded to do it on my own little typewriter—with two fingers! That report's abominable typing was disregarded, and reactions to its contents were—by academic standards—lightning fast and strong.

Then a letter arrived, inviting me to teach economics at Harvard. Letter in hand, I rushed to see my manager, Ralph Gomory. Very pleased, he sent me to his manager, Herman Goldstine (1913–2004), director of mathematical sciences at that time. I had met him in Princeton when he was John von Neumann's second in command; because I had been relatively unimportant at IBM, this was—since Princeton—the first real contact we had.

"What brings you here today?" "I would like to be granted a one-year leave of absence to teach at a university." "You know that this department strongly believes in supporting teaching. I'd be delighted to grant this request, and please remember that we cover the difference between the IBM and university salaries." "By the way, you left me no time to mention that I'm invited to teach economics." "Oh! I would have expected statistics or applied mathematics, but this is fine. More important, you haven't named the university that is inviting you?" "Oh, I am very sorry. I should have started by saying so. The invitation is from Harvard, and their offer is higher than my current salary." At this point in our conversation, he became extremely agitated and reached for a pillbox.

Harvard Called and IBM Noticed Me

If status within IBM had been measurable, mine would have instantly jumped from well under the radar to well above—where it stayed. This jump was far more important than the accompanying raise in salary from well below the norm to slightly above.

Leaves from IBM became part of an ongoing arrangement. That IBM granted academic leaves was a novelty, and outsiders watched without a clue as industrial labs came and went, and often questioned my position's durability. I wondered as well, but reasoned that a huge and fast-growing computer manufacturer genuinely needed frontier research.

I long put out of mind the fact that IBM failed to provide the university guarantee of tenure. But not for a moment did I forget that to remain stable and vertical, a bicycle must move sufficiently fast. In pedantic words, I made a clear distinction between the static stability that tenure would have provided and a dynamic form of stability that was unexpected, makeshift, and constantly affected by shifts in both IBM and the outside world.

From the day I informed Herman Goldstine of my first "call" from Harvard, two things started to become increasingly clear. Invitations from elite outside institutions strengthened my position inside, and my attractiveness to outsiders could not be taken for granted. I became acutely sensitive to the question, "What have you done lately for our science?" My record as an innovator needed to be constantly enhanced by conspicuous fresh achievements. Each visit to a corner of academia contributed in ways that neither IBM nor academia's other corners could possibly match. But to stoke my "innovation furnace" at IBM was best.

As a result, times at IBM and times in Cambridge and elsewhere came to alternate, an order slowly emerged, and distinct works were reshaped as aspects of a whole—a fractal geometry of roughness. The sequence that followed, of sociological low points and intellectual highs, could not have been predicted.

How I Came to Study Price Variation

Let me stop to tell how I became fascinated with price variation, a completely new topic for me. The background resides in some work I had done earlier on an ancient topic—the law of distribution of personal income, which had been discovered in the 1890s by Vilfredo Pareto (1848–1923). That law—and my work—intrigued a few economists, and I was invited to speak at a Harvard seminar directed by Hendrik S. Houthakker (1924–2008).

Upon entering "Hank's" office, I got a surprise that made that day one of the most memorable of my life. A peculiar diagram on his blackboard seemed to me nearly identical to one I was about to draw in my lecture! How was it, I soon asked, that something I had just discovered about personal incomes was already on display? "I have no idea what you are talking about. This diagram concerns cotton prices." He had been working with a student before I arrived, and the blackboard had not yet been erased.

Why should the way income or wealth is spread throughout society relate to the ups and downs of the price of cotton? Why should both cases exhibit the same pattern of concavity and convexity? Could this reveal a deeper connection between these two aspects of economics—some odd truth lurking behind the charts? By then, mainstream writers on finance had rediscovered the old theory that prices vary as if by the toss of a coin. They were looking for evidence, but reliable historical records were hard to come by. Cotton was an exception.

For more than a century, the New York Cotton Exchange kept exacting daily records of prices as the vital commodity moved from the plantations of the Old South to the dark mills of the industrial North. Virtually all interstate trading was centralized at one exchange. This should have been an economist's dream, but to Houthakker and his student, it proved a nightmare. There were far too many big price jumps and falls. And the volatility kept shifting over time. Some years prices were stable, other years wild. "We've done all we can to make sense of these cotton prices. Everything changes, nothing is constant. This is a mess of the worst kind." Nothing could make the data fit the existing statistical model, originally proposed in 1900, which assumed that each day's price change was independent of the last and followed the mildly random pattern predicted by the bell curve.

In short order, we made a deal: he'd let me see what I could do. He handed me cardboard boxes of computer punch cards recording the data. "Good luck. If you can make any sense of them, please tell me what you find."

Back at IBM, the computing center assigned a programmer to analyze those records—as it had for income distributions. How many big price jumps, how many small? The wait being long because I was low on the priority list, I took the train to Manhattan, where the National Bureau of Economic Research was then located. Its library included many dust-covered books filled with tables of financial data—a treasure before the time of computers. Later, I obtained records from the U.S. Department of Agriculture in Washington, D.C. Gathering every available piece of data, I built an encyclopedia of cotton prices, daily, weekly, monthly, and annually, over more than a century.

What the computer helped me find was extraordinary. Houthakker's view was confirmed: the price changes from one day to the next, one week or month or year to the next, did not behave as the 1900 model assumed. The variance misbehaved. Each time I added a price change to the data set, my estimate of the variance changed. It never settled down to one simple number of volatility. Instead, it roamed erratically. That was surprising, considering that the quality of the data itself could not be challenged. Moreover, there were too many big price jumps to fit the bell curve.

Two Pictures of Price Variation

How do prices vary on the organized markets called bourses, stock exchanges, or commodity exchanges?

For centuries, such markets have thrived without the benefit of a systematic mathematical model. The first such model was put forward in 1900 by an outsider in French mathematics, Louis Bachelier (1870–1946). It came out astonishingly early—well ahead of its time—and was odd indeed. It became the standard financial model, and was the one Houthakker was using with cotton prices. The model was advanced financially, but not buttressed by any data whatsoever. Originally, it drew little attention, but over time two events revived it. On the theoretical side, Norbert Wiener rediscovered it around 1920 as a model of an important phenomenon called Brownian motion, which became more developed in the 1930s and '40s. On the concrete side, the advent of the computer in the 1960s empowered the study of both the data and the theory.

Thanks to the computer, I was able to note the flaws in Bachelier's model in a rough report I wrote in 1962, and put forward a counter theory that could be stated with no formula, which truly fit my wild Keplerian dream. And it produced in 1963 my first paper on this work, "The Variation of Certain Speculative Prices," which was to be frequently cited in the economics literature. The 1900 theory assumed that price jumps can be neglected—the mathematical concept being "prices vary continuously"—and that price changes follow the same rules in prosperity or depression. Much well-documented evidence contradicting this theory made frequent and large ad hoc "fixes" necessary. My 1962 counter theory allowed for discontinuities and was later extended in 1965 and 1973 to allow for alternating periods of prosperity and recession.

All price charts look alike. Sure, some go up, some down. But daily, monthly, annually, there is no big difference in their overall look. Strip off the dates and price markers and you cannot tell which is which. They are all equally wiggly. "Wiggly" is hardly a scientific term—and until I developed fractal geometry years later, there was no good way to quantify so vague a notion. But that is exactly what we can now see in the cotton data: a fractal pattern. Here, the fractal scaling up and down is not being done to a shape, such as the florets of a cauliflower. Rather, it is being applied to a different sort of pattern, the way prices vary. The very heart of finance is fractal. So it all comes full circle. It was no coincidence that Houthakker's cotton chart looked like my income chart. The math was the same.

Unfortunately, my careful tests that should have blown up Bachelier in 1963 failed. The economics profession decided that my work was too complicated and too unfamiliar. The departure it represented and further threatened was hard to develop and sell. It seemed far easier to continue with an endless stream of "fixes." What was I to do? I moved to an altogether different set of "priority interests," with only episodic returns to price variation. The 1900 theory of finance that I had discredited has persisted, attracting many young mathematicians and scientists, depleting the fields they come from.

And then, perhaps a bit later than I expected—in 2008—the market did what it was bound to do: it crashed.

Kepler Versus Ptolemy

Bachelier's 1900 and my 1963 models of price variation were the first two to be put forward and are the stars of the events that will unfold shortly. Is the topic doomed to be presented forever in terms of this contrast? I am afraid it is and—in all modesty—would like to explain why, by comparison with a key event in science: the replacement of the ancients' faulty model of the motion of planets by Kepler's ellipses. Ptolemy's model stated that the planets revolved in a circular orbit around Earth. However, he regularly had to amend it when he observed anomalies. This belief was widely held until the 1600s, when Kepler proved that the planets revolved around the sun in an elliptical orbit.

Bachelier had assumed that price changes follow an ancient frequency curve—a familiar one—called Gaussian. Its key property is that large deviations from the norm have an absolutely tiny probability and therefore do not matter. In the following price fluctuation charts, the top chart is Bachelier's model, showing that most changes

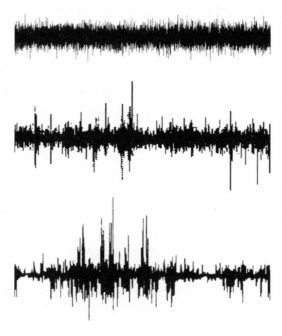

are small. The middle is a real price chart for IBM's stock price, show-ing some outliers and far greater fluctuation than Bachelier's model. The bottom chart uses my computer-generated multifractal model, showing that it stands up to actual records of changes in financial prices.

I provided the data with an intellectual home—not surprisingly, one that hardly any practical person knew of at that time. Indeed, Paul Lévy's teachings had familiarized me with a curio he called sta-bility and I prefer to call Lévy stability. Hence, I could identify the behavior as a characteristic of price change. Working at IBM, I had access to a computer center where—laboriously—Lévy stable densi-ties could actually be calculated for the first time.

In the case of cotton—with no fiddling—the fit was striking. My first work in finance had brought together two domains of knowl-edge far removed from one another. Surprisingly, the stable distribu-tion fit every detail of the data—in particular, a symmetry in the distribution that earlier examinations had missed.

It is often asserted that if one adds many statistical quantities, even the largest is conveniently negligible in comparison to the sum. How-ever, the contrary had long been known to happen—but only in cases with which practical statisticians need not be concerned. A loud oppo-nent of mine repeatedly claimed that those cases were "improper," but this view led him astray. In fact, those cases were well known to experts, but were viewed as belonging to irrelevant pure mathemat-ics. By bringing them into all-too-practical finance, I imposed and argued for a deep distinction between "mild" and "wild" states of ran-domness of chance. To the best of my knowledge, all past work on prices did not conceive of this wildness and had confidently relied on the reality being ruled by randomness that was "proper," therefore mild.

The three states of chance—wild, mild, and slow—can be com-pared to the three states of matter. Are not solid and gas separated by liquid? Absolutely. In my view, the same is true of chance—the coun-terpart of liquid being "slow" randomness. And liquids happen to be enormously more complicated to study.

A Backhanded Compliment?

There can be little doubt that Mandelbrot's hypotheses are the most revolutionary development in the theory of speculative prices since Bachelier's initial work [of 1900. His] *papers force us to face up in a substantial way to those uncomfortable empirical observations that there is little doubt most of us have had to sweep under the carpet up to now. With determination and passion he has marshaled as an integral part of his argument evidence of a more complicated and much more disturbing view of the economic world than economists have hitherto endorsed. . . . Mandelbrot, like Prime Minister Winston Churchill before him, promises us not utopia, but blood, sweat, toil and tears. If* [he] *is right, almost all our statistical tools* [and] *past econometric work* [are] *meaningless. . . .*

Surely, before consigning centuries of work to the ash pile, we should like to have some assurance that all our work is truly useless.

This mixture of faint praise and insidious attack is from a text by the economist Paul H. Cootner (1930–78). I first saw it in December 1962. Technically, this attack would have been easy to answer, but politically, this and others like it led me to conclude that business economists' blind loyalty to Bachelier's 1900 theory was too ingrained to be overcome. So I simply stepped aside. No matter what I might say or do, heavy criticism was bound to follow. Leaving French academia for an American industrial laboratory—a colossal gamble—had proved I was prepared to take controversial stands. At that time, however, the last thing I could afford was to be a Mr. No. So I swallowed hard and moved on. Soon enough, I would tackle another, less political problem: turbulence in fluids and its extension to large scales, commonly called the weather.

Financial orthodoxy is founded on two critical assumptions in Bachelier's key model: price changes are statistically independent, and they are normally distributed. The facts, as I vehemently argued in the 1960s and many economists now acknowledge, show otherwise. First, price changes are not independent of each other. Research over the past few decades, by me and then by others, reveals that many financial price series have a "memory" of sorts. If prices take a big leap up or down now, there is a measurably greater likelihood that they will move just as violently the next day. It is not a well-behaved, predictable pattern of the kind economists prefer—not, say, the periodic up-and-down procession from boom to bust with which textbooks trace the standard business cycle. As my later work showed, it is a more complex, long-term memory—one that can be analyzed fractally. Second, the distribution of price changes is not "normal." Conventional theory says that if you measure the changes from one day, hour, or month to the next, the vast majority of the changes should be very small, with only a few days exhibiting big changes—the "outliers" on the standard bell curve typically used to graph them. In fact, there are many more big changes than standard theory says there should be—many more days when prices crash or soar.

Before sweeping all that "uncomfortable" data under the carpet, Cootner should have examined how much information he was destroying. This was hard to illustrate in the 1960s, but now it has become very easy. Next to the plot of an actual price index, test what would happen, at every moment, if Cootner had "swept under the

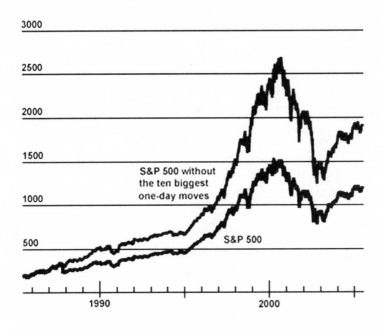

carpet" the x largest price drops over x days. The index would more or less double. In other words, the few largest increments that Cootner discarded and the increments he preserved are equally important. The figure above dramatically illustrates this. The bottom line plots the actual daily S&P 500 price index from 1990 to 2005, and the top plots the same data without the ten largest daily price moves.

A second key misstep of Bachelier was even more serious, and correcting it took years. All along, everybody who cared about price variation knew about business cycles. The analysis I carried out in 1962–63 mixed together the phases of low and high price variability. A more realistic model must dig deeper into the data. A common way out introduces the concept of business cycles and assumes that different phases of a cycle follow different rules. Unfortunately, cycle timing has always been mysterious, unreliable, and discernible only long after the fact.

The first challenges to Bachelier were entirely my work, and for decades continued to develop largely in my hands. In due time, I was able to claim that variations in financial prices can be accounted for by a model derived from my work in fractal geometry. An observer cannot tell which of the data concerns prices that change from week to week, day to day, or hour to hour. This quality defines the price charts as self-affine fractal curves and makes available many powerful tools of mathematical analysis. Fractals—or their later elaboration, multifractals—do not claim to predict the future with certainty. But they do create a more realistic picture of market risks than does observation alone. Eventually, and most unexpectedly, I combined that work with my older theory of word frequencies, and this led me to the fractal geometry of roughness.

Does My 1963 Model Always Apply?

My model does not always apply. In my first printed article on price variation, "The Variation of Certain Speculative Prices," I underlined the meaning of "certain" by pointing out areas where the issue was not closed. In later life, I was often asked to talk about prices. Once, when my key interest was the alternance in fluids between laminar and turbulent zones, I was showing a trusty old slide on prices when—quite suddenly—I perceived an uncanny resemblance to a feature of turbulence called intermittence. Not a completely new idea, perhaps, insofar as the banker John Pierpont Morgan (1837–1913) had reputedly claimed the market was as fickle as the weather. While lecturing on, I told myself, How silly you have been. Don't mention your thought today, but make sure to look into it when you can. In due time, I did.

Will Training a Grad Student in Finance Lead to a Job in Chicago?

Eugene F. Fama, a student at the University of Chicago Graduate School of Business, often visited me. More important, I met his Chicago adviser, Merton Miller (1923–2000), who convinced his business school colleagues to hire me. First, he brought me to Chicago, with Aliette. Although there was heavy snow all around, my lecture was mobbed—and a big party followed, given by the dean, George Shultz.

It was clear that an offer had already been arranged and Chicago simply wanted to see what they were purchasing. At some point in my visit, my total lack of experience with U.S. universities led me to take a characteristically costly and foolish step. A written offer could always be "traded" for advantage in some other place—at IBM, in the University of Chicago, or elsewhere—but I had no mentor. Talking with Shultz made me realize that all they knew about me was an IBM report. They had not even looked up my vita and so did not know that in addition to being a freshly anointed pioneer in finance, I had extensive other interests. Standing at this critical fork in the road, I unwisely enlightened them. Shultz was very warm and commented that it was wonderful that one salary was going to get them several different professors.

Back in Boston, the telephone rang: George Shultz on the line. He thanked me for that beautiful lecture, et cetera, et cetera, then came to the point. The offer was withdrawn. Really? He had rushed to ask several other departments whether they would share my salary. The answer was always no—they did not even know me. This left him with a problem that was the bane of my life: my tendency to cross scientific disciplines. He feared that my interests might move out of economics as smoothly and unexpectedly as they had moved in. This represented a risk he would not take.

He was disappointed that his diplomatic skills had not been sufficient. He also reassured me that my thinking in economics would be well represented, because Eugene Fama was going to join the faculty.

There is irony in this. This was the same Fama who, in 1964, submitted a thesis subtitled "A Test of Mandelbrot's Stable Paretian Hypothesis." He believed that successive price changes were statistically independent. I had to convince him that I had never claimed independence and that he was in fact testing a much weaker hypothesis—the one that was first expounded in Bachelier's 1900 Ph.D. thesis and had become known as the martingale hypothesis. Fama conceded, corrected his earlier assertions, replaced the mysterious label "martingale" with "efficient market," and built his career on becoming its champion. This hypothesis is convenient indeed, and it is, on occasion, useful as a first approximation or illustration. But on more careful examination, it failed to be verified—and for being its herald, Fama should receive neither blame nor credit.

Continuing to follow my lead, he supervised several excellent Ph.D. dissertations. Next he returned to the fold and had a brilliant career as one of the leaders of his profession's backlash to the strictest Bachelier orthodoxy—suitably "nostrified" by a new vocabulary. Naturally, the University of Chicago soon stopped inviting me.

Shultz's path and mine crossed again once at a gathering of U.S. residents holding the Légion d'Honneur. He remembered the episode and thought it had ended well. Probably the diplomat was being diplomatic. Shultz went on to run Bechtel, a huge construction corporation in California, then became, successively, Nixon's director of the Office of Management and Budget, secretary of labor, and secretary of the Treasury. Later, as President Reagan's secretary of state, he brought his diplomatic skills to the world stage.

23

On to Fractals: Through IBM, Harvard, MIT, and Yale via Economics, Engineering, Mathematics, and Physics, 1963–64

YES, I HAVE BEEN WARNED. This chapter's title seems to make no sense. How can it possibly reflect reality? Surprisingly, it does approximate this period of my life. All too often while waiting to deliver a lecture, I've heard the host conclude the customary introduction by wondering how I could conceivably exist. In fact, this chapter's title is only a raw outline. Numerous additional fields I visited also differ deeply yet share a key feature that to me matters more than any other: roughness.

Instead of the gritty term "engineering," why don't I use the more genteel "applied science"? In part, because one word is better than two. But largely because I want to make a point. The phenomena I have studied are elusive and not yet covered by any proper quantitative science—pure or applied. Think of a distant past. Water mills came long before the applied science of fluid mechanics; heat engines came before the applied science of heat. Stock exchanges arose before any theory, and no theory existed to which my work on finance could be "applied." My ambition was more realistic—that is, more limited—yet essential. I wanted to provide a consistently more faithful description of known facts—and hence help financial engineering out of its dismal and harmful state. The same goes for the developments that will be described in this chapter: no existing body of science could assist them.

What I have just said explains why I did not fear moving into a variety of problems of engineering. To master many applied sciences would be an idle dream—especially for an outsider like me—and is a process that would be unwise to rush.

I led a complicated life in parallel to my quiet one at IBM—a life as a teacher or researcher who meandered from place to place or from one field to another. The fact that my life's most productive season came late kept me in a constant hurry, and I could rarely take it easy. Before the fact, the path I was following seemed wild and impossible to manage; after the fact, it seems unavoidable. Halfway through each discovery—no sooner—I experienced a marvelous and exhilarating surprise.

Had I come to IBM either when it was not yet ready or when it was already all too well organized, I might have returned to French academia. Few of my colleagues had my luck with an unplanned and truly extraordinary coincidence: that of an individual ready to experiment across many different fields and a corporation willing to trust that individual's judgment. Essentially, all I had learned through my otherwise too long and too scattered wandering years gradually changed from a random burden on my memory to a very valuable asset in my work.

I worked in the four fields of research in this chapter's title—economics, engineering, mathematics, and physics—and I became involved with fractals in the arts. The first three are the departments I taught in as a visiting professor at Harvard. Of those three, nothing beats my impact on finance and mathematics. Physics—which I fear was least affected—rewarded my work most handsomely. My other work influenced rather small communities. A deep unity that had been present in my work all along was gradually revealed, then increased its presence and became my guide.

Having worked in many fields but never wholly belonging to any, I consider myself an outlier. It does not hurt that the word "outlier" has an established technical meaning in statistics: it is an observation that is so very different from the norm that it may be due to accidental foreign contamination. A classic example concerns astronomical observations that were contaminated by cats residing in the observatory.

Yes, cats walking across the observatory floor shook the telescope a bit, causing some orbits to be miscalculated. For two centuries, economists and statisticians have looked for good ways of preserving real data while eliminating would-be cats. To the contrary, I have found that the so-called outliers are essential in finance. In fact, a common thread of my work is that values far from the norm are the key to the underlying phenomenon. In many fields I remained far from the norm, so in that way too I am an outlier.

Hydrology: The Biblical Joseph, Hurst, and Me

There come seven years of great plenty throughout all the land of Egypt: And there shall arise after them seven years of famine. These words are found in the Bible. Look up Genesis 41:29–30 for the story of how Pharaoh had a dream and his high helper, Joseph, son of Jacob, interpreted that dream—and saved Egypt from a famine by storing enough grain for the lean years.

An indomitable self-taught scholar who went on to graduate from Oxford, hydrologist Harold Edwin Hurst (1880–1978) interpreted that dream as reflecting a feature of the notorious variability of the waters of the Nile. Known as Abu Nil (Father of the Nile), Hurst remade himself into an expert on the Nile River Basin and became a champion of the need for the Aswan High Dam. He spent years searching modern data for a "signature" of Joseph's interpretations. The topic being hot and potentially very costly, many experts were called in. In 1951, Hurst proposed a solution for optimum dam design based on what his research found. The experts opined that this formula by an undereducated author could not possibly hold.

In a 1965 publication, I showed that while Hurst had no clue about what he had discovered, his formula indeed holds—and has unexpectedly far-reaching consequences. To a scientist it means that the span of dependence in the flow of the Nile is infinite—while for the Rhine it is finite, and even short. What a joy to quote the Bible as a (pure) scientific reference! But did all that matter in practice? I heard that the Aswan Dam's engineers, instead of following Hurst, had followed the cold war international political imperatives.

The study of rivers brought me to the distinction between two kinds of fractals: the self-similar (shapes scaled by the same amount in every direction, like coastlines) and the self-affine (shapes scaled by different amounts in different directions, such as turbulence).

My explanation of Hurst's formula was another Kepler moment. After it was published, I pursued the mathematical aspects with mathematician John W. Van Ness. Then I wrote a long series of papers with hydrologist James R. Wallis. IBM Research prided itself on bringing the two of us together. It seems that many big dams are built in China. I wonder whether they are Hurst-Mandelbrots.

Distribution of Galaxies

That the Milky Way is one of a number of similar "objects" in the sky is a surprisingly recent notion: it dates from the decade when I was born. So is the galaxy. Incredible but true, however, long before any evidence became available, the concepts of a galaxy and clusters of galaxies had been repeatedly invented and forgotten. Also, the natural assumption that faraway shining objects are uniformly distributed in space was analyzed and shown to lead to the embarrassing Olbers paradox, which argues that the sky must be uniformly and infinitely bright. A way to avoid this paradox was proposed by a science fiction writer, Edmund Fournier d'Albe, and developed by astronomer Carl Charlier. But the profession never took it seriously, largely because it required the universe to have a well-defined "cluster" and because relativity theory demands a well-defined overall density of mass. Somehow, I found out about this bit of esoterica, instantly identified Fournier d'Albe's model as a primitive fractal, and proposed one, then another, less primitive models.

The title of a draft of my first paper on galaxy clusters implied that clustering is an illusion. In duller words, this is the way data are spontaneously interpreted by the human eye, but not necessarily a property of the problem at hand. "Tell me if I understand correctly what you have been telling us. We astronomers take it for granted that galaxy clusters are real things out there," the man said, pointing his finger to the sky (well, the ceiling). "What you propose is that they may very well be right here." He pointed to his temple. "Is that right?"

Sometime around 1990, I was in a Tyrolean resort hosting a meeting on the large-scale structure of the universe. The questioner was a person I did not recognize and never saw again. I felt very good. At long last, a key aspect of my fractal model of galactic intermittence—one I had patiently described in each of my "Fractals" essays—was receiving a reasonably serious hearing.

The audience might have cursed me for being a troublemaker: I was bringing new tools to a corner of astronomy that had been placid; I was sowing doubt and creating new problems. Observers used to take what they saw for granted: galaxies are organized into clusters, which are themselves organized into superclusters—a splendid new use, as I saw it, of Ptolemy's classic model of the motion of planets. For the "reductionists"—theoreticians whose business it is to "reduce" everything to a field's basic principles—the task is to explain why galaxies cluster and in doing so to predict the cluster sizes.

My alternative to Ptolemy is far more parsimonious and suitably Keplerian: I claim that the distribution of galaxies is fractal. The point is that in some fractals, clusters are completely real because they have been included by construction; in other fractals, no clusters have been included by construction but the mind sees them anyhow. Fractality and hierarchy manifest a peculiar consonance. Below are two images of galaxies: on the left, a real galaxy cluster from the Center for Astrophysics and Space Sciences at the University of California, San Diego, and, on the right, a computer fractal model of galaxies.

My analysis led me to conclude that up to a certain depth in the uni-

verse, galaxies had not a uniform, but a fractal, distribution and were easy to construct. With half a line of formula, I got all this clustering of galaxies—superclustering—out of it. That is, my model automatically reduced the overwhelming complication of reality to a single basic principle—a principle at the core of science, one which tries to duplicate the complication of reality by using very simple rules.

Helping Lady Luck Through Telephones

Do you recall that my testing of cotton prices began with a mysterious diagram on a blackboard? Well, Lady Luck struck again when I was asked to help with some troublesome noise on data transmittal telephones, and I found a way I liked of thriving as a jack-of-all-trades.

An odd thing is that chance has helped me on many occasions. Louis Pasteur is credited with the observation that chance favors the prepared mind. I think that my long string of lucky breaks can be credited to my always paying attention. I look at funny things and never hesitate to ask questions. Most people would not have noticed the dirty blackboard or looked at the article that Szolem pulled from his wastebasket for me to read.

That 1951 reprint and that diagram on the blackboard are both examples of what are now called long-tailed or fat-tailed distributions. These episodes made me the first well-trained mathematician to take those tails seriously. As a result, I have sometimes been called the father of long tails. Whether long or fat, those tails are an intimate part of the fractal family. So it makes perfect sense that I have since been called the father of fractals.

In the spring of 1962, when my friends at IBM heard about my going to Harvard, they pushed me to give a seminar—just to explain how someone like me had managed to land "this plum" (their words). I obliged, convinced them that I could not help them manage their savings, and found myself in the flattering yet burdensome position of a man who could work miracles . . . and perhaps assist them in their IBM jobs. The questions I was asked were unpromising, except for one that looked interesting, but was a long shot. Linking computers to telephone lines was proving harder than expected, and a friend at IBM, Jay M. Berger, had been assigned a problem concerning the distribution of errors on those lines. He and his assistants were supposed to find out why they clustered this way. None of the textbook laws of averages seemed to apply. To paraphrase Julius Caesar, I came, saw, and was immediately hooked. Once again, I brought together a problem from one world and a tool from a far-removed other world. A second major Kepler moment within a year.

The reports from Berger's group satisfied his managers. A paper I wrote with him on error clustering in telephone circuits provoked a tempest in a small but very important teapot. The experts caught on and soon my work became standard material. I was invited to the epicenter of expertise, Bell Laboratories. In time, they stopped sending me their papers, and I stopped asking them questions. But the seed had been planted.

Of Galileo's many gifts to scientific knowledge, here is an essential one that requires no formula. His world believed that the heavens were orderly, while everything on Earth was a mess. To the contrary, Galileo found plentiful messes on the moon—its craters. He also found order on Earth—the falling of stones pulled down by gravity. In this sense, George Kingsley Zipf—whom we met when I told the story of my Ph.D. thesis—was solidly pre-Galilean. He believed that in the physical sciences, randomness follows the distribution called normal, Gaussian, or bell-shaped, while in the social sciences—word frequencies, personal income—the distribution is the so-called hyperbolic.

★ ★ ★

Finance was far from filling my time during the academic year 1962–63. Every week—in addition to teaching economics—I was compensating for academic deprivation at the bucolic IBM Research. When not asked, I shamelessly volunteered to speak at one seminar or another at Harvard, MIT, or elsewhere. Also, I managed to attend innumerable seminars on countless topics—a form of continuing education that my overworked local friends could only dream of. The substance of my talks, prepared at IBM, concerned the first (Pareto-Lévy-Mandelbrot) of my three successively improved models of financial prices. The creative aspect involved new input that triggered late work in hydrology and the second (Hölder-Hurst-Mandelbrot) model in finance. My Harvard years in applied sciences were a direct outcome of Hölder-Hurst-Mandelbrot, but soon involved another new input that led to work on turbulence and the third financial model.

To an incredible degree, the incessant wild motion of that year has left a deep trace throughout my life. My schedule was so packed that my self-inflicted wounds at the University of Chicago soon began to heal.

After one repeat of my standard talk on price variation to a group of noneconomists, an auditor—to whom I am greatly indebted—spoke up. He observed that some aspects of what I said reminded him vaguely of something he had heard about the variability observed in the discharges of rivers. I became very excited. This was about the time when the Berger-Mandelbrot paper on telephone errors was published. Also, economics had led me to worry about oil fields. So I knew that two examples of scaling in the physical world had to be added to Pareto's law of income distribution and my work on prices. The river discharges promised to add a third, extremely different one, so I rushed to visit the Harvard hydrologist Harold Thomas. He referred me to the work of the hydrologist Harold E. Hurst. Solving the Hurst puzzle tested my skills but took little time.

Snatched Up by Harvard Applied Sciences

I had no ongoing contact with the people who were involved in this episode of my life, so my memory of what followed has faded. First, I was asked to return to Harvard the next year and give a few lectures on my findings. I did not find this appealing. Then I was introduced to a physicist who was the dean of applied sciences. He proposed to bring me back to Harvard in 1964–65—but a move back to IBM and another to Cambridge would have been a logistical nightmare and did not please Aliette.

We settled on an alternative: continue at Harvard in 1963–64, but take a few steps north, from economics to applied sciences. A few steps across Harvard changed the environment. In economics, when I asked for stationery, the chairman's office provided a starter supply and added that I would surely want to use my own letterhead. In applied sciences, there was a proper stationery office, open stack, like at IBM. My economics office had its own telephone. In applied sciences, I shared a telephone line with three or four regular professors, including a Nobel Prize winner. An additional visitor who did not mind hogging the phone forced them to add a new shared telephone.

We rented the house of noted MIT physicist Victor Weisskopf, who was on leave directing CERN in Geneva. The attic contained piles of French comic books like *Tintin*. Aliette read them to young Laurent, then suggested he look at them by himself. He did and in the process learned to read French. Later, Didier followed the same path.

Weisskopf was a charming and cultured man. When I saw him last—in Alpbach in his native Austria—he was eighty-four. At a festival, a band of vacationers heard me lecture on mathematics and then him on physics. During the discussion, I pointed out that my lecture was presented as an idiosyncratic view of my field, and his as a general talk on physics. At lunch, he complained about how hard it was for him to finish his memoir, and urged me not to write mine too early—certainly not as long as I still could do science. I promised, and now can only hope that my wait has not been too long.

Teaching at Harvard Applied Sciences

A one-term course on the Hurst puzzle of persistence in hydrology was mentioned in private conversations with the dean. But the public announcement listed a perennial title, Topics in Applied Mathematics, adding that the instructor for the 1963 fall term was going to be me.

The first day repeated my feat of the fall of 1962. An exceptionally large class had assembled. The reason the dean was so agreeable to my coming became clear: the division offered too few courses.

Second surprise: not one hydrologist attended, so I wrote a brief note in French to claim credit for a breakthrough but put developing it on hold—where it stayed for several years. That large class had assembled because of my paper with Berger on errors in telephone chan-

nels. Some other students had exhausted Harvard's slim pickings in electrical engineering. And there were a number of postdocs and senior researchers.

My material on telephone errors was a bit skimpy for one term, but a good soldier (or good actor, if you prefer) does not say, "I can't." So at many class meetings, I reported on progress since the preceding meeting—often made the morning before the class. Terrifying stimulation, but very effective. Only once was I forced to sneak into the classroom minutes ahead to write on the blackboard, "Due to unforeseen circumstances, today's meeting is canceled." On another occasion, I started by asking the class to forget the previous week's two meetings because ten-minute-long substitutes I had developed over the weekend were easier and also went further.

One student, a naval officer, had to cut short an assignment at Harvard and report in a few months for submarine duty. To satisfy the rules, he needed one more credit but had already taken every other Harvard offering he might consider. Though unprepared, he asked to be accepted as a charity case. I agreed and reassured him that visiting lecturers were not supposed to fail anybody.

Later, when he brought his term paper, he asked for permission to make a brief statement. "Sure, go ahead!" "Sir, as I had told you, I was absolutely unprepared for your course. My paper is not good at all—

it's all right. Whatever grade you give me will not affect my career. The main thing I wanted to say is that in your course I did learn something very important. I had been told that science was created by humans, but in all my other courses it seemed created by creaky machines. Your course made me watch science being created. Thank you, sir. It was a great experience, sir. Good-bye, sir." With that, he clicked his heels and exited my life.

Extremely moved and convinced he was not faking, I thought of my uncle Szolem's visceral dislike of "elegant lecturers." They make everything seem crystal clear, but looking at your notes in the evening, you realize that some detail had been forgotten and without that detail everything collapses. Szolem preferred—and practiced himself—the very different style of the man who taught him French mathematical analysis in Kharkov during the civil war that followed the Bolshevik Revolution. Serge Bernstein corrected himself constantly, as if reinventing or at least fully appreciating for the first time the material he was teaching. Szolem would say, "As he went on, he seemed to painfully tear mathematics from his body."

The year 1963–64 marked my crossing back from the social to the physical sciences. Trifling issues known only by a few specialists and understood by nobody—so that they were called "anomalous" or went under many other empty names—led me into the midst of a key scientific topic: intermittence in turbulence.

No Permanent Position at Harvard

Friends apparently took it for granted that Harvard would offer me a permanent position in applied sciences in a context broader than had Chicago. Aliette and I fondly hoped that would be the case. Rumors grew, then stopped. I inquired and found that I had indeed been considered, but my overly optimistic friends had not weighed in as firmly as needed. One likely opponent was the eminent expert in fluid mechanics, George Carrier (1918–2002). After I explained to him an early form of my multifractal description of turbulence, he responded that if that direction were to prevail, the study of turbulence would cease to interest him.

Ultimately, my interests and achievements were viewed in Chicago as absurdly broad, and at Harvard as absurdly narrow! Unfortunately, I had to agree that those opinions were not entirely unreasonable. I did not fulfill Chicago's specific needs but was readying to move through many other sciences.

In Voltaire's *Candide,* the ever-optimistic Dr. Pangloss claims that everything turns out for the best in the best of all possible worlds. Given that my fate was to conceive and develop fractal geometry during the following ten years, Pangloss could argue that neither Chicago nor Harvard would have provided me with the right environment. This tired argument would praise torture as a way of enhancing sainthood. And as Harvard concluded, the swath I was about to cut was to be comparatively narrow. Not expected by them was its being ubiquitous, highly visible, and widely influential.

After 1964, I stopped worrying about the suitability of IBM and set to work. What I accomplished during the decade mirabilis of the sixties was to culminate with an annus mirabilis at Harvard in 1979–80.

A Rare Institute Lecturer at MIT

Harvard was out, but—as was said—Aliette and I had grown enchanted with life in Cambridge. In hindsight, remaining there would not have been the best choice, but at that time we would have much preferred to stay. Therefore, I took the short path to MIT worn by generations of Harvard rejects and refreshed an old relationship with Jerome Wiesner. Jerry had sailed with John F. Kennedy and had been his widely acclaimed science adviser, a post he kept briefly under Lyndon Johnson. He had returned to MIT and was at that point dean of science.

In no time, he arranged for me to become Institute Professor at MIT. Peter Elias (1923–2001), an old friend of mine and Jerry's successor as head of electrical engineering, took care of the paperwork. That this should have been possible was a tribute to Jerry's skills and the level of institutional flexibility he maintained. Institute Professorships were originally meant for scholars crossing many fields, but they soon began to be granted to former administrators or became the most senior chairs in traditional departments.

Jerry was an extremely low-pressure salesman. "Here is an offer, but don't take it. President Kennedy was from Boston and all-out for Cambridge, so President Johnson is all-out against it. Everybody fears major funding trouble. Several departments want you to be here, but each expects somebody else to pay. You would be the only Institute Professor without a deep and strong constituency; therefore, for you, safe funding will become increasingly hard. More generally, funding for science is becoming threatened. Believe me: for somebody like you, MIT is all wrong and the right place is IBM. Manny Piore wants people like you and is far freer to move than any university. Don't take my offer." With clenched teeth, I followed Jerry's advice.

Next Jerry recalled a splendid precedent. MIT had Visiting Institute Professors: Arthur Kantrowitz and Polaroid's Edwin H. Land (1909–91), then at the height of his fame as an inventor, scientist, and one of the richest men in the world. Neither had an office on campus, and their appointments were open-ended, with no term.

Marvelous? No, too good to be true. The possibility was killed by a better-informed Mr. No higher up. He pointed out that giving those titles to Land and Kantrowitz had drawn sharp fire from a group of activists—including Aliette's cousin Leon Trilling. Existing Visiting Institute Professorships were allowed to continue, but new ones were out of the question.

Worn out, we watered that glorious title down to Visiting Institute Lecturer. IBM readily agreed, and for many years this compromise remained a wonderful and fruitful arrangement. I made fairly regular one-week visits to dizzyingly varied groups at MIT and elsewhere in the Boston area. More-or-less chance encounters—often like my first meeting with Houthakker—continued in a steady flow and provided an extraordinary supply of new thoughts and new directions that could be instantly explored within IBM Research.

To summarize and conclude, Chicago, Harvard, and MIT had honored me by trying to bring me in—but didn't. The best manners were shown by gritty MIT, followed by upstart Chicago.

I contributed to each conclusion by being a truly dismal politician who preferred working to networking. However, the mismatch that was repeatedly demonstrated between academia and me was genuine. I had not a single identifying brand name for my activity. Ten more years went by until I gave up and coined the word "fractal." Unlike me, my linguist friend Noam Chomsky had his MIT career smoothed or oiled by an attractive and assertive flag and several brilliant friends who rode along with him and found support.

The lack of choice was frustrating—but there is absolutely no question that on my return to IBM I churned out a mass of work, much of which had a rapid impact.

Lady Luck Against the Mess of Turbulence

In his *Odyssey*, Homer relates the problem that Ulysses encountered while sailing the long distances from Troy to his home in Ithaca—and everywhere between Scylla and Charybdis. Today those trips would not be scary at all, but in Ulysses' day, boats were not built to fight the unpredictable turbulent weather encountered on long voyages. The problem of turbulence is so hard that every small step forward is a reason for pride. A fellow visitor at Harvard was Robert Stewart from Vancouver. He was an expert on turbulence. In one of his seminars, he analyzed records taken by a decommissioned submarine he monitored as it was slowly moving near Vancouver collecting data. Both in space and in time, the turbulence in the ocean it sailed through proved to continually come and go—by "intermittence," as they said. During Stewart's talk, I sat in the first row, smiling from ear to ear, rejoicing at the great gift I saw arriving. The work I was then doing on noisy channels—the next step after my 1963 paper with Jay Berger—fit Stewart's data wonderfully and could use the same techniques. My feat in connecting the thread from an engineer's headache to reputedly wild mathematical esoterica was not a fluke!

For years, straining to understand turbulence a little better was one of my favorite means of self-mortification. I became familiar with yet another set of experts I had not known and soon ceased following. My papers added up to books on this topic.

In 1964, when I returned to IBM, I realized that the Hausdorff dimension I had learned first from Henry McKean at Princeton and later from Paul Lévy was ready to move from esoterica to reality. In the context of prices, the measurement of volatility was the Hausdorff dimension. In the context of turbulence, the dimension of roughness was again the Hausdorff dimension.

I developed a multifractal model that addressed the intermittence of turbulence and has also turned out to be fundamental to our understanding of the variation of financial prices. Qualitative properties like the overall behavior of prices, and many quantitative properties as well, can be obtained by using multifractals at an extraordinary small cost in assumptions.

After my two years at Harvard, IBM corporate headquarters wanted me to take a job at Cornell University. This was tempting, but Cornell is in Ithaca, in upstate New York. Aliette and I had often visited, but we feared isolation and decided to return to Yorktown.

Excellent decision. I experienced the warm feeling of coming home to the delights of old-fashioned collegiality in a community far more open and "academic" than Harvard. The cafeteria had no competitors nearby, and even home-cooked lunches were eaten there. I loved particularly the varied conversation at the so-called physicists' table—which was of course open to all comers. The mathematicians' table was smaller, more homogeneous, and far less disputatious. The physicists and friends exchanged news—more often focused on science and scientists, and also on music and history, than on local or national politics. And frankly, nowhere else could I find a more diverse and appreciative audience for my stories.

24

Based at IBM, Moving from Place to Place and Field to Field, 1964–79

THE TIME BETWEEN MY FIRST ARRIVAL at Harvard and the publication of *The Fractal Geometry of Nature* stands out as my life's middle period. It began exceptionally late, so I continually felt in a great hurry and ranged in directions far more varied than I would have thought sensible or feasible.

Did I have a firm research agenda? Only in my head and mostly in a form that—whenever needed—could instantly be erased, shuffled around, or changed. To a degree that others would have found intolerable, I rarely managed to do what I was dying to do. Instead, I was doing what happened to be most desirable given what I perceived as the market for scientific ideas like mine—or, in other cases, what I viewed as easiest to undertake given some special resource that had become available in one corner or another of a very large institution.

Most fortunately for me—and for science—the physicist Richard Voss joined IBM in 1975. A freshly minted Berkeley Ph.D., he came in large part at my urging and became an essential ally and a close friend. He is a creative free spirit with extremely broad interests, and a true master of the computer. Other associates—bringing with them some specific skill—came and went; most stayed for a year or two.

Trumbull Lecturer and Visiting Professor of Applied Mathematics at Yale

Back when I was in Paris, working at Philips and writing my dissertation, a statistician named Leonard "Jimmie" Savage (1917–71) was there on a sabbatical. He had been at the University of Chicago during the "Stone age," when the unquestioned boss was Marshall Stone. Next he moved to Michigan, then Yale. I respected him greatly for his fortitude (he was nearly blind) and the breadth of his reading. For example, he alerted American academia to the 1900 Ph.D. thesis of Bachelier. But our actual interests had little overlap. Although we never became close friends, we kept in touch and saw each other on my rather frequent visits to Yale.

Harvard's old Lawrence Scientific School had a (less well-endowed) Yale counterpart, the Sheffield Scientific School. One of its buildings had been given to the math department, and the others were being continually reorganized. It had openings, and Savage suggested that it might be a good place for me. So I came to be tested. The 1970 spring term began with three packed Trumbull Lectures and continued with a short course—referred to as a seminar—on my various models of "abnormality" in the real world. It was well attended, and a few people identified themselves as being deprived of such activities. No offer came. Anyhow, I had lost interest.

To my shame, the overall stillness of Yale—contrasted with the incessant goings-on at MIT—created the impression that not much was happening. I did change my mind, but only seventeen years later.

In Paris: A Lecture Not to Be Forgotten

On January 16, 1973, I lectured at the Collège de France in Paris—an occasion that no attendee could forget. This was a very special event because, as long as Szolem was a professor there, his fear of condoning even the slightest possible appearance of nepotism was rare, extreme, and irrational. Only after he had retired could his former colleagues think of inviting me—which they did promptly.

I spoke at an interdisciplinary seminar that two senior professors organized on Saturday mornings. André Lichnerowicz (1915–98), a professor of mathematical physics, was famed for his broad curiosity, good taste, and political skills. He was scheduled to be on my Ph.D. committee, but illness had prevented it. François Perroux (1903–87) was a professor of economics. When the announcement was being drawn up for posting on bulletin boards, Perroux commented that working at IBM was undignified; any academic affiliation was far preferable. The Harvard economics department would have been great, but I was no longer there. A nominal and unpaid affiliation with the National Bureau of Economic Research was deemed satisfactory.

Preparing for this seminar turned out to be a major undertaking. It forced me to gather all I had achieved and fit it into an hour. This effort started me on my 1975 book. Invitations were sent to several luminaries in Paris, and word of mouth helped spread the news. As a result, the medium-size auditorium where I spoke was absolutely full. The talk itself was fairly general—a summary of topics I had worked on. But the discussion that followed brought out wide-ranging and very precise questions. I answered each, briefly but technically. In a sense, I gave a dozen five-minute technical presentations. As the meeting proceeded, my homecoming was palpably turning into a coming-out in the Paris big leagues—a rare major event. The discussion continued in the courtyard. A friend commented that he had never heard a strictly scientific lecture that was also so blatantly autobiographical.

A few days later, *Le Figaro,* a major daily, published a big column by one of those who attended and spoke, Pierre Massé (1898–1987), to whom I had been introduced. Under Charles de Gaulle, he had been a celebrated commissioner for planning. Before that, he ran the state electricity board, and early on he was one of the brilliant engineers who had built hydroelectric dams all along every suitable river in France. His endorsement may have precipitated the episode to which I shall now proceed.

Deciding Not to Compete for the Collège de France

A question arose: Would returning to Paris be either desirable or manageable? Gradually, it became clear—to put it simply—that it was not.

This decision was soon tested by a totally unexpected telephone call from André Lichnerowicz. "Your talks at the college have left a continuing and very favorable impression. François Perroux has now retired and his chair is open. Candidates are plentiful but none are impressive. Several of us would prefer you. If you express strong interest, you will be elected."

High praise and a credible guarantee. I knew from Szolem that in Collège de France elections prominence outside one's field generated unexpected enmities as well as support. I had only two substantial accomplishments, both very technical and undeveloped—a fraction of my present total. Marvelous to hear, this was enough to gain support and open up a unique and splendid occasion to return in the highest possible style.

I had come a very long way. The promise I had shown at those old examinations in 1944–45 had never been forgotten in Paris and was being given a chance to be fulfilled.

"You must be warned that the Collège de France is *une auberge espagnole*," Lichnerowicz continued, invoking a tired old ethnic slur that implied that to eat and sleep in Spanish inns one had to bring one's own food and bedding. "Organizing a group to help you may take much time and effort. Please think about it and call me back."

I felt like Julius Caesar before he crossed the Rubicon to conquer Rome. The institutional forces that made me leave France in 1958 remained entrenched and invincible—but I would return in a much stronger position, perhaps sufficient to keep those forces at bay. Moreover, the Collège de France shared with IBM Research a feature not present elsewhere in academia. While I would be elected on the basis of my work in finance, I could teach anything I chose. This mattered to few persons other than me.

I realize now that I was about to be pushed out of the economic mainstream by a major step in academic economics: the 1972 revival by Black-Scholes-Merton of the formula of Louis Bachelier. Could I have both fought and outwaited them in a protected site?

Unfortunately, the downside was big. From the viewpoint of the dream that ruled my life, the timing was dreadful. Fractal geometry was on a roll, and at IBM I had squirreled away sufficient resources to prepare the 1975 French book and undertake a longer English one. Furnishing a "Spanish inn" properly would delay or perhaps even kill those plans by opening me up to the temptations that a Collège de France chair presents to an opinionated intellectual in Paris.

What I am about to say may sound ridiculous. Burning scientific ambition came first, and I would not think of endangering it. I might have considered compromising, since the yearly duties of a Collège de France professor easily fit into one term. This allowed Szolem and others who had neither a laboratory nor a growing family to spend every second term in the United States. IBM might be agreeable. In fact, that triumphal lecture in January 1973 must have contributed to my becoming an IBM Fellow a year later, making me much freer. A better politician less subject to jet lag might perhaps have accepted and continued part-time at IBM. But my position was delicate. I did not want that job enough to accept the offer tendered by Lichnerowicz; I thanked him but turned him down. I couldn't tell if he was surprised.

Mother Dies in 1973

Paid sabbaticals were not officially offered at IBM but could be negotiated. I was scheduled to be in Paris as a Guggenheim Fellow during the year 1968–69. But political upheavals—the events of May 1968— intervened. There was no doubt that they would be followed by a bad hangover, during which visitors would be unwelcome, or at least uncomfortable, especially those with a Parisian background. Therefore, my sabbatical was postponed until 1972–73, when Mother's health became preoccupying.

Only slightly younger than Father, Mother aged quite well. She took care of Léon's three daughters and followed my career with swelling pride but no active influence. When Léon moved to his current flat, she moved to a smaller one in the same building. Her physician, a brilliant and colorful cousin of Aliette, admired and loved her and thought it was a good idea to send her each summer to a "cure." Her letters to the local doctors were firm: Mother was simply old and should be given a very mild regime. One year, the local doctor was clumsy, so Mother went to a different doctor whose treatment was so vigorous that Léon had to bring her back home. Her energy had been very diminished. Aliette and I offered to relieve Léon from the burden of caring for Mother during the summer of 1971, and IBM gave me a sabbatical to spend the year 1972–73 in Paris. We were close when she deteriorated and died in January 1973. Her life had been long and endlessly complicated—but ultimately fulfilled and happy. One of the last times I saw her (barely) alive, I described a great event: my lecture at the Collège de France. I hope she heard and still had the strength to rejoice.

Visits to the Mittag-Leffler Institute

The mathematics research institute of the Royal Swedish Academy of Sciences, located in Stockholm's elegant suburb of Djursholm, occupies the former mansion of the colorful Victorian Gösta Mittag-Leffler (1846–1927). His wife's Finnish forests allowed him to build a mansion to his taste. It consists of a large but not extravagant bourgeois apartment on the main floor, with several former servants' quarters. A three-floor library that would make any university proud houses many valuable old books and much space for collecting. Mittag-Leffler wanted to teach mathematics without having to move to the Swedish Oxbridge, so with a few friends in Stockholm, he simply endowed the most private university imaginable, seed of the present university of Stockholm. He also created his own journal, *Acta Mathematica*.

The Mittag-Leffler Institute restricts itself to its namesake's field of mathematical analysis. Each year (sometimes each term), it tackles a different topic, and its glory years were those under the lay directorship of Lennart Carleson, particularly the ones when his dynasty included his frequent coauthor, Peter Jones. A topic has to be selected several years ahead. I was thrilled that three of the topics chosen over the years were from my work. The Mandelbrot set was selected in 1984, before it became the height of fashion, with the hope of solving the Mandelbrot Locally Connected conjecture; a big effort ensued but failed . . . to this day. My 4/3 conjecture about Brownian motion was chosen in 1998, when its difficulty had become obvious and it seemed that a solution would be hastened if all those concerned could be brought together. As it happened, the solution came before the meeting, so the meeting was able to draw immediate consequences. The third meeting, in 2002, that my work inspired was on the mathematics of the Internet.

As you may have experienced, some non-negligible proportion of e-mail gets lost. Multiple identical messages are a pest, but the sender is actually playing it safe for the good reason that in engineering everything is finite. There is a very complicated way in which messages are assembled, separated, and sorted. Although computer memory is no longer expensive, there's always a buffer of finite size somewhere. When a big piece of news breaks, everybody sends a message to everybody else, and the buffer fills. So what happens to the messages? They're gone—just flow into the river.

At first the experts thought they could use an old theory developed in the 1920s for telephone networks. But as the Internet expanded, it was found that this model would not work. Next they tried one of my inventions from the mid-1960s, and it wouldn't work either. Then they tried multifractals, a mathematical construction I had introduced in the late 1960s and into the 1970s. Multifractals are the sort of concept that might have been created by mathematicians for the pleasure of doing mathematics, but in fact it originated in my study of turbulence. To test new Internet equipment, one examines its performance under multifractal variability. This is even a fairly big business, from what I understand.

25

Annus Mirabilis at Harvard: The Mandelbrot Set and Other Forays into Pure Mathematics, 1979–80

WHAT I ACHIEVED OR STARTED during the spring of 1980 went well beyond the wildest dreams of my adolescence under foreign occupation.

"I see you carry another batch of computer pictures. Are they your latest? May I have a look? Hmm . . . To me, they tell absolutely nothing. How can one possibly extract any kind of mathematics from such squiggles? Can this game really concern that ancient theory of Pierre Fatou and Gaston Julia on the iteration of rational functions? Their time seems long past."

When and where did I first hear comments and questions like

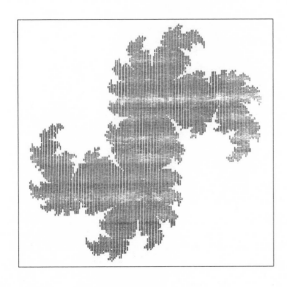

those, and what motivated them? It is true that I have heard many such utterances throughout my life, but I heard them with special intensity during the 1980 spring term whenever I approached the pigeonholes that held the professors' incoming mail.

It was my third yearlong visit to Harvard, but my first in the mathematics department—a visit quite different from the earlier ones. A new world was being revealed to mathematicians . . . or perhaps an older world was being painfully revived.

Day after day, colleagues, students, and passersby witnessed a slowly unfolding process—one I had never lived through, and one the Harvard community of pure mathematics had not experienced for generations and did not in the least expect. For me, the process was intoxicating. For the mathematicians, it was baffling at best, and in many cases unwelcome or worse. This process was a step-by-step transmutation that began with almost meaningless ink smudges that were transformed first into rough observations, then into increasingly more precise ones, and finally—insofar as I am concerned—into fully phrased mathematical conjectures. The resulting pictures were amazing.

These pictures were intriguing objects I then called lambda and mu-ma—alternative ways of representing a fundamental new mathe-

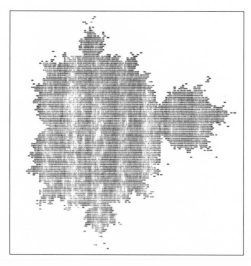

At first there was one "island," then more. As these "offshore" islands began to appear, they were hard to differentiate from specks of dirt.

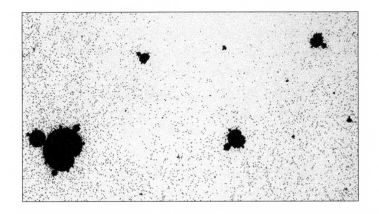

matical structure that became known as the Mandelbrot set. It has been called the most complex object in mathematics, has become a topic of folklore, and remains my best and most widely known contribution to knowledge.

I could only prove the simplest conjectures. I knew I would be unable to prove the harder ones, so I had to abandon them, complaining all the while and loudly calling for a full and rigorous proof. Skilled mathematicians at Harvard and in Paris were informed and soon gathered; shortly after, they proved several of my conjectures and many of their own. As decades passed, numerous additional conjectures joined mine, and many have been proved in exquisite ways. My first key conjecture has been rephrased, yet has survived multiple expert searches for proof and remains proudly open.

Today—thirty years after those heady events—the branch of mathematics that my conjectures revived continues to shine brightly.

A Luncheon That Changed a Life

How did all this come to be? In the mid-1970s, I often saw Stephen Jay Gould (1941–2002), a lively paleontologist with multiple appointments at Harvard. Quite independently, we had become two very visible champions of discontinuity—he in paleontology and I in the variation of financial prices. Early in 1977, I was visiting Boston and called him to see if he was free for lunch while I was in town. He was, and a date was set.

He came with a mathematician friend from Harvard, the number theorist Barry Mazur. Barry often visited Paris, was fluent in French, and had read my 1975 *Les objets fractals* with enthusiasm. I showed him the brand-new, expanded English version, the 1977 *Fractals*. A spirited conversation was cut off by our respective schedules. It was a Friday, and Barry invited me to brunch at his house the next day. Who would refuse?

At brunch, he pressed me on two topics. One was the original papers and books of the early days of real analysis, the period around 1900 when it was viewed as a collection of diverse mathematical "pathologies"—toys, in my thinking. The second topic was the already substantial number of cases where I had transmuted such a toy into a tool. As we talked, Barry said, "You know, this would make a wonderful course in our department. The current course in real analysis has become so fast moving and streamlined that the concepts seem to come from out of the blue, unconnected to any motivation. I had thought of a supplemental course to fill in the gaps, but I don't know the history that well and could not come up with even one real application. No one else could. Would you be interested in trying?"

Indeed, I was! Arranging for me to be invited, Barry could not have imagined what he was setting in motion. However, my younger son, Didier (a toddler on my previous visits), was to be a high school senior in 1978–79, so we could not leave that year. Instead, we settled on 1979–80. As it turned out, Didier went on to Harvard, so we all moved to Cambridge together.

As the fall of 1979 approached, the contract for my leave from IBM was not yet finalized. But I was told not to worry, because during the fall term I would just be visiting Harvard privately and would not teach until the following term, in the spring of 1980.

Physics in Broken Dimension

Having time to pursue research that fall allowed me to begin what became a long-term collaboration with Amnon Aharony, a physicist from Tel Aviv University who had been visiting IBM in Yorktown.

I listened to several of the talks he gave. After one of them, I made some comments. "You know what," he responded, "you may be right. That crazy mathematical idea of shapes of fractional dimension may well have a useful bearing on my kind of physics. We must work together and take a close look."

So we did and became deeply involved, first at Harvard during that year of miracles, and later in many places over many years. We investigated the use of shapes of fractional dimension. Most of our papers concerned spaces where dimension is not 1, 2, 3, or higher but a fraction, and brought fractals toward the mainstream of statistical physics.

This peculiar notion entered mathematics and physics independently, and each discipline responded differently. Mathematicians offered many definitions, while physicists proceeded heuristically, essentially asking if the calculation had predicted what was observed— the proof being in the pudding. In this collaboration, everybody's skills were essential, and the result "smelled good"—though it was not final. This led me to put forward a bold conjecture: that solving the usual partial differential equations of physics can yield either familiar and expected smoothness, or fractality.

A Grand Old Problem Frozen in Time

How did the Mandelbrot set arise and provoke such strong reaction? Basically, from a challenge that I "inherited" from Uncle Szolem when I was a student in the 1940s.

"One of the oldest, simplest, and greatest problems in all of pure mathematics reached a peak decades ago with Pierre Fatou and your teacher Gaston Julia. Then—for lack of new questions—their work screeched to a halt. It must not be allowed to remain frozen. I tried myself, very hard, several times, to revive it, but always failed. Over a quarter century, everyone who tried also failed. Go and see what you can do. Here are reprints of their old papers. Hold on to them because they are rare and quite valuable."

I took Szolem's advice and the reprints, hoping to report back that I went, saw, and won. I went, but saw nothing I could advance. Just as Szolem and everyone else had, I looked for questions combining sufficient novelty with a sufficiently good chance of being answered. I failed.

Early in life, I learned that for a scholar, nirvana is to take an unsolved problem that had been stated long before and solve it. I also learned that a mathematical problem could be well stated yet remain unsolved for a long time, even centuries, while a whole field develops around it. And I understood from readings and course material that a field might simply die for lack of manageable and interesting unsolved questions. All this brought faint solace.

As I see it now, the thoroughness of the failure to further the work of Julia and Fatou implied that the missing ingredient could not simply have been a more powerful or ingenious angle on mathematics as it existed in their time. For one thing, between the 1910s and 1950, power had drifted to the friends of André Weil, the members of Bourbaki, who quite deliberately focused on entirely different problems.

Then, in 1953–54, when I was at the Institute for Advanced Study in Princeton, another key ingredient began to take hold. My sponsor, John von Neumann, was trying to interest his colleagues in the equations of the weather. Only a handful of mathematical equations—already known in the eighteenth century—had solutions given by explicit mathematical formulas. In all other cases, including those for von Neumann's study of the weather, no such solutions were worth dreaming of. To Johnny, this indicated that one should seek answers from numerical simulations using the computer. But Johnny died before convincing anybody.

An overwhelming majority of the mathematicians of that day shuddered at the very thought that a machine might defile the pristine "purity" of their field and deliberately erase the past. Starting with my work on prices, I immediately understood the power of the computer, even though I never learned to program one. When I was at Harvard, a colleague reported being astonished that a computer could help one of the Ph.D. candidates achieve control of a difficult mathematical problem. In no way did any mathematician expect that a great ancient problem could be revived by the computer. A revival does not happen all by itself—never has. In this case, it would happen because of a combination of chance events in my life.

One was a lengthy obituary of Henri Poincaré written by Jacques Hadamard, Szolem's predecessor and sponsor at the Collège de

France. After Hadamard died, Szolem put together his collected works, and he gave me a set. The obit dwelled on a dry-sounding mathematical topic called limit sets of Kleinian groups, which I knew about from books for advanced high school students. I was reminded of the time Szolem persuaded me to revive Julia and Fatou, and my interest was sparked once again. Having a Princeton math student as a "visiting assistant" for the holidays, I sought and formed a construction of Kleinian group limit sets.

A Turning Point in Mathematics

This story hinges on a very plain formula—the only one to be allowed in this memoir. "Allowed" does not mean having to be understood, appreciated, or acted on. It suffices to observe that the formula is very short:

> Pick a constant c and let the original z be at the origin of the plane; replace z by z times z; add the constant c; repeat.

In mathematical notation, this instruction would reduce to three letters and three symbols. In mathematical lingo, this is a quadratic map, something close to an ancient curve called a parabola. But in the Mandelbrot set, z denotes a point in the plane, and the formula expresses how a point's position at some instant in time defines its position at the next instant. Again, in mathematical lingo, this formula defines the very simplest form of dynamics in discrete time—a form called quadratic dynamics. Fine, the formula is indeed breathtakingly simple. So why bother?

This formula is then iterated—that is, repeated with no end—defining with increasing refinement a shape that can be approximated using a very simple computer program.

To wide surprise, this shape is both overwhelmingly rich in detail and minutely subtle, and it continues to provide a common and fertile ground for exploration: from Brahman mathematicians to students and those in the earthy lower castes, from artists to the merely curious. Its infinite beauty, appreciated by so many, was completely unexpected and brought forth countless challenges, which mathematics and philosophy have not yet exhausted. Immediately and with no prompting, zooming in on a point on this set's boundary fascinates all eyes, young and old.

Needless to say, I don't feel I "invented" the Mandelbrot set: like all of mathematics, it has always been there, but a peculiar life orbit made me the right person at the right place at the right time to be the first to inspect this object, to begin to ask many questions about it, and to conjecture many answers. Though it had not been seen before, I had a very strong feeling that it existed but remained hidden because nobody had the insight to identify it.

Zoom toward a point on the circumference of a circle and the curvature gradually "irons out," yielding an increasingly straight line. But zoom instead toward a boundary point on the Mandelbrot set and what you see becomes ever more beautiful, wild, baroque, and complex in many distinct ways, which the set of color images in this book illustrate. I have heard it described as "pretty—yet pretty useless." Important applications of new discoveries take time to be revealed, and we have seen that the Mandelbrot set has powerful redeeming features. Thought wanders to Napoléon's saying that a good sketch, in all its complexity, is worth a thousand words, or even to the biblical *Let there be light*. By now, I should have become blasé, but I hope I never will. With due humility, these magic words of Charles Darwin apply:

> From so simple a beginning,
> endless forms most beautiful and most wonderful
> have been, and are being, evolved.

More or less actively, I have lived with this set for over thirty years—and would have been thrilled to live with it far longer, were it not that success invites too many other seekers.

Preview of the Mandelbrot Set at the New York Academy of Sciences

The mathematical theory of chaos was a hot topic in the late 1970s and the focus of a big conference on nonlinear dynamics held at the New York Academy of Sciences in 1979. I had not yet discovered the Mandelbrot set, but I spoke about my work on iteration as it stood just before that key period and could not resist giving an idea of the slide show spectacular that I was beginning to carry around the world.

The audience was overwhelmed, and there were few questions. But there was a follow-up not to be forgotten. That was the day's last session, and—to my delighted surprise—my IBM colleague and friend Martin Gutzwiller asked me to show those pictures again. Most of the audience stayed for the encore.

The proceedings of that conference became a major reference book. When the time came to turn in my section, I submitted instead the first announcement of the key facts about the Mandelbrot set. The announcement included several of those early pictures. Worried that the printer would think the pictures were ink smudges, I added this instruction: "Do *not* clean off the dust specks. These are real and important."

I gave the same talk at Harvard, where a form of mathematical physics was a topic of broad and growing interest. The last of many

questions came from David Mumford (b. 1937), an algebraic geome-
ter and laureate of the Fields Medal, a professor of mathematics at
Harvard, a colleague, and a warm host I wish to thank here. "Couldn't
the same approach devise a fast algorithm for Kleinian limit sets?" For
a hundred years, this had been a goal of many mathematicians,
including great ones, and probably also of countless amateurs. Astonishing but true (and, given the simplicity of the "trophy," perhaps
almost embarrassing), all those seekers had failed.

David asked if I could look at Kleinian limit sets. Delighted, I
responded that—at least for one important special case—I had indeed
discovered a construction and showed him the draft of my paper. He
marveled, then observed that the tools I had used were ancient,
utterly elementary, and certainly intimately familiar to Poincaré,
Robert Fricke, and Felix Klein—skilled men who had first raised the
question a hundred years earlier. The self-inverse limit set on the left
led me to the more general Kleinian group limit set on the right.

 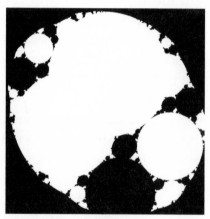

David wondered aloud what made me succeed where those seekers and so many others had failed. My answer distilled—once again—
the already told story of my scientific life: when I seek, I look, look,
look, and play with pictures. One look at a picture is like one reading
on a scientific instrument. One is never enough.

At that point in history, Kleinian groups were in a holding pattern.
Towering figures—including Lars Ahlfors (1907–96) at Harvard and
Lipman Bers (1914–93) at Columbia—had made great strides. But the

impression prevailed that their act was hard to follow, and hardly anyone was interested in my algorithm. But one day, the overly thin walls of my Harvard office allowed me to overhear the words "Kleinian group." The speaker, who turned out to be a colleague, S. J. Patterson, confirmed that there was little interest in the topic. I convinced him that this perceived lack of interest deserved to be tested, and a seminar was organized. We had about thirty people at the first meeting!

Mumford naturally attended, and he became very supportive of my work. In record time, my assistants taught him computer programming. I also introduced him to David Wright, the student I overheard talking with Patterson. Mumford admitted that he held summer jobs as a programmer but thought that, as a Harvard graduate student in mathematics, he should not advertise this heresy. I assured him that before long it might cease to be one. Indeed, many mathematicians—though surely not all!—soon became enthusiastic about the power of the computer, and Mumford moved away from algebraic geometry, a field in which he was a major figure. He experimented on the computer with Kleinian groups richer in structure than those I had looked at. One of his earliest illustrations, implemented on a visit to IBM, first appeared in *The Fractal Geometry of Nature*. His interests have now moved on to a computer-based theory of vision.

Zigzagging Through the First Course Ever on Fractals

The first course on the topic of fractals—which I improvised from day to day in the spring term of 1980—was closely related to my ongoing research. The course was not for credit, and the auditors ranged from young undergraduates to seasoned Ph.D. candidates.

To assist the course, I wanted live demonstrations and was advised to hire a senior, Peter Moldave, who turned out to be one of the best programming helpers I ever had. The personal computer had not yet hit the world, and none of the students had a clue about how to use one or about the benefits of its graphics. This, along with Peter's prowess as programmer, created a sensation. Moreover, Peter was not taking any hard courses during his final term and said he would be delighted to assist me with my research.

The course limped along until after the spring break, when it morphed into something entirely different: the first-ever public discussion of my discovery of the object that—very late that year—came to be called the Mandelbrot set. Peter's help was essential to that discovery.

★ ★ ★

Ultimately, Harvard did not work out. I was expected to pursue and teach my style of using computers, but computers and their use were not welcome at Harvard. Hence, there was a near-total absence of both equipment and skills among the students and faculty. And because personal computers had not yet become ubiquitous, those who absolutely needed them went elsewhere or had private and well-hidden facilities.

Having been so outspoken about the sad state of computing at Harvard, I was disappointed that good news did not come until it was too late for me. Out of the blue, David Mumford informed me that the National Science Foundation was responding to critiques like mine by establishing a supercomputer at Geometry Center at the University of Minnesota.

Wide Wonder, Complexity, and Mystery

The Mandelbrot set strongly appeals to three very different groups to which I belong: those interested in pictures, complexity, and pure mathematics.

Pictures

The complication of the actual pictures obtained when that little quadratic map is repeated very many times—starting with $z = 0$—is overwhelming. Of course, one could not repeat the formula by hand, only by using a computer. The earliest and "rawest" of these pictures were in black and white—or, more precisely, darkish and whitish grays. The constants are not ordinary (real) numbers, each attached to a point on the line, but complex numbers, each attached to a point in the plane.

Millions of examples can now be found in hundreds of books and on the Web.

What about the colors? The defining formula yields a whole number: 1, 2, 3, and so on. To cut down on incomprehensible clutter, I replaced ranges of numbers with shades of gray. Then colors took over. Selected by the programmers, they were completely arbitrary and a reflection of good or bad taste.

Complexity

When I set out to study that rule that ends with the word "repeat," I decided with little reason that nothing of much interest could possibly come from such a simple map.

Around that time, Andrei Kolmogorov and my IBM colleague Gregory Chaitin had, independently of each other, attempted to measure the complexity of a mathematical structure. They put forward the length of the shortest sentence that could implement that structure. Where does this position the Mandelbrot set? Is it the most complex set in the whole of mathematics, as some have asserted, or is it as simple as its generating formula? I could not decide and concluded that the question begs to be restated in a different way. But given the stark discontinuous contrast between an input and an output that today is nearly instantaneous for the Mandelbrot set, many view it as extremely—miraculously!—complex. I feel exceptionally privileged that my wanderer's life led me to be the agent of this discovery.

Pure mathematics

Had anybody investigated "that set" before I did? No, nobody had. After the fact, extraordinary efforts were made to find predecessors. A claim was put forward on an unmotivated drawing that was too crude to show anything but had been appended to a paper—without comment. Also, someone read through one of Fatou's long papers and found a mention of "that set" among related ones, but without further discussion or anticipation of any result.

To my surprise and profound delight, my original paper on the subject was an absolute first. The title is "Fractal Aspects of the Iteration of [Quadratic Maps] for Complex [Parameter and Variable]." It appeared in late 1980 in *Annals of the New York Academy of Sciences*.

Inevitable question: Was *Annals* the worthiest place to publish a groundbreaking paper? Not in the least. But I was embarking on a lecture tour, and a printed text was urgently needed. So I replaced an expository paper I had read at that academy with the more recent work and, wherever I went, carried copies of the proofs. Did this work? This is the paper that led to "that set" being named after me, yet in the early days—when it mattered—hardly anyone quoted it.

* * *

So, ironically, my best-known discovery did not result from the availability of exceptionally good pictures at IBM. It was made at Harvard, where I had to deal with complicated research conditions within a very bad system. The pictures we saw on the first night seemed incomprehensible; the second night, they became more coherent. Within a few days, they had grown completely familiar, as though one had always seen them. Incredible!

How does the importance of the Mandelbrot set compare to that of fractal finance, which is highly influential in a well-defined community of "practical people"? All my diverse "children of the mind" are equally dear to my heart; they can't and shouldn't be compared. In that case, what makes me perceive 1979–80 as an annus mirabilis? My work in 1962–63 made for a wonderful year, but it was a year of a single miracle that developed slowly over time, while the 1979–80 miracle came on like lightning—as miracles should.

26

A Word and a Book: "Fractal" and
The Fractal Geometry of Nature

NEVER UNDERESTIMATE THE POWER of a word that appears at the right time and in the right context and—let us not forget—accompanied by the right pictures. The word "fractal" has spread like wildfire to so many minds, books, and dictionaries that it is hard to believe it dates only to 1975. The underlying idea had been written about every so often since time immemorial, and a skeptic may wonder if it was really necessary to invent a word to describe my work.

How did this word, "fractal," come into usage? I had to coin it when the French edition of my book was being written—the need for a word had become convincing and I had become confident that it would fly. Did I act like a superstitious parent who names a child only after its birth? Let it be. I also checked in advance that "fractalist" would sound good if a need were to arise for a word to denote me and the followers I hoped to inspire.

Like my speech in every language, my scientific writing in every field carries a strong foreign accent. Because of this anomaly, several of my papers were rejected and other drafts did not seem worth finishing. Instead, they got filed in some dark corner of my personal archive.

As a result, a backlog of unfinished drafts began to grow until, at one point, my friend Mark Kac volunteered some unexpected but truly excellent advice. "Most active young scientists know they must publish articles or perish. But your case is different. Unless you stop this avalanche of individual articles and write a book, I shall let you perish." I am extremely grateful to Mark for this "command."

I solved my communications dilemma by publishing a great deal of original work in three books. They arose as successive versions of a

broad-based "essay" combining a fractal manifesto and a casebook—
that is (using military terms I don't like but find hard to avoid), a call
to arms and stories of successful past campaigns.

The 1975 "Preview" Book, *Les objets fractals*

When my soon-to-come-out book was still tentatively titled, in
French, *Concrete Objects of Fractional Dimension*, the publisher, Flammarion, was horrified and asked for something better. Friends con-
curred. "You have written about a brand-new idea. You are entitled—
in fact, obliged—to give it any name you want. Make it snappy."

I could have given a new meaning to some already overloaded old
word (think "catastrophe" or "chaos"). But I chose to coin a new
word—one not directly evocative of anything in the past. I wanted to
convey the idea of a broken stone, something irregular and fragmented. Studying Latin as a youngster taught me that it is a very concrete language. My son Laurent's Latin dictionary confirmed that the
adjective *fractus* means "broken" or "shattered." From this adjective, I
thought of the word "fractal."

We scrounged around IBM and put together tools to produce a
camera-ready manuscript. I had been "managing" the book from
cover to cover with a tiny "staff": one constantly changing full-time
programmer (and French typist) and one or two part-time associates.
This was a powerful moment of triumph against seemingly overwhelming odds. An exhilarating experience for everyone involved,
pushing me to the limit and demanding enormous effort from everyone on the project.

No book is published without some expectation of success, but for
the original French preview, the chances of success could not conceivably be forecast. Flammarion had agreed to take the risk of publishing
the book only because I had been introduced to the boss by a mutual
friend. Sales were slow at first but after a while picked up nicely, and
the fourth revised edition is in print today as a popular pocket book.
Many years after the first publication, several French mathematicians
confided that my book had a great influence on them when they were
students. In 1975, however, this bright fate was far in the dim future.

My slim volume was made part of an illustrious series that had at one point published Henri Poincaré, Jean Perrin, and Louis de Broglie. In 1975, it was barely alive, but it seems that my book revived it.

When I offered a copy to Szolem, he first congratulated me nicely, then thumbed through it and, seeing it was not a math book, asked, a bit testily, "But what kind of book is it? For whom have you been writing?" My answer: "I don't know but hope it will create a readership for itself, perhaps even a large one." My cousin Jacques was present; amused, he asked his father, "In your case, when you write a book, you always know exactly who is going to read it, right?" Szolem responded, "Yes, there are about fifteen people in the world who read everything I write. That is enough. I find that very comforting."

A tiny event comes to mind. New books in French were few, but bookstores were numerous and prominent—many occupying locations now taken by travel agents and off-price stores. I came to know personally a lot of the hands-on owners or managers. The manager of the bookstore Offilib was a friend since the time I helped him settle down. He took me aside for a piece of advice: "Your book is marvelous, enchants everyone. But watch out: don't let yourself be carried away and spend the rest of your life trying to improve it. Go back to something standard and build yourself a reputation that will ease your career." Advice that—of course—I failed to follow. That bookstore sold so many copies of my 1982 English book that the projected French translation was called off.

The 1982 Book, *The Fractal Geometry of Nature*

Going to Harvard in 1979, I carried with me the computer tape of what I perceived to be the nearly ready third version of the book. But because that year turned out to be an annus mirabilis, the shadowy third version kept being expanded to mention the Mandelbrot set and the first mainstream physics papers. It also kept being reorganized in response to what I learned by teaching that first course on fractals in 1980. Ultimately, I started almost from scratch, and the much-expanded text went smoothly. I succeeded in persuading W. H. Freeman's top brass to charge a low price for the book *and* include a sixteen-page color signature (at a time when color was still perceived as expensive) because I felt it would be a good investment. And it was. As feared, the book ran late, but the color signature was available and I took it around to meetings.

First Fractals Meeting in Courchevel

In July 1982, while waiting for copies of *The Fractal Geometry of Nature* to be shipped, I had the pleasure of delivering its content to representatives of the scientific world and watching their reaction. The occasion was the first fractals meeting ever held. The venue was Courchevel, an exclusive ski resort high in the French Alps. The page proofs of *The Fractal Geometry of Nature* had reached me in Paris, where I spent hours at IBM preparing a photocopy for every participant. From Paris I lugged rock-heavy suitcases. The audience numbered about fifty and was very heterogeneous and not representative of any group, since fractals had no constituency.

At this meeting, the first I ever organized, nobody could really help me. Almost every author of a contribution to fractals was invited to speak, and IBM branches in different European countries sent a few people whose goodwill seemed valuable. Half of the slots on the program remained empty, and I put my name in each, hoping that somebody in the audience would relieve me from this commitment. Miraculously, the down-to-earth Summer Institutes held before and after mine motivated IBM Europe to provide a computer of good size, by the standards of the day. In addition, my IBM colleagues and close friends Richard Voss and Alan Norton had come along.

In the absence of core organizers, I wrote very few follow-up and reminder letters, and my travel instructions were complete but without frills. Many participants later confided that they were not quite sure that the meeting was actually going to take place, and reaching the conference hotel gave them a strong sense of accomplishment. Because none of the home institutions of the participants could afford such a machine and such skilled help, the computer room was filled until well past midnight.

In the summer, ski resorts close or charge little, and the management promised that the hotel would be empty. However, when I arrived a few days before the meeting, the manager expressed profound apologies. The European Youth Orchestra had begged him to rent the vacant half of the hotel, and he had no choice but to agree. He assured me that the musicians would make good neighbors: they would be working so hard that nights would be very, very quiet. Besides, the orchestra promised to allow us to attend the general rehearsal before its job-seeking tour of the music festivals. Also, I would meet its leaders: the Chicago Symphony Orchestra conductor Georg Solti (who could not spend much time visiting with conference attendees because the high altitude sickened him) and Claudio Abbado, the future director of the Berlin Philharmonic. Illustrious past and shining future—not bad at all! When I opened the meeting on Sunday night, I could brag that the usual musical divertissement would be kindly provided by my friends, the maestros Solti and Abbado, live. Of course nobody believed me, but they realized at the concert that I had not been pulling their leg.

I chaired the entire meeting and channeled the discussion vigorously. Also, even though I eventually found other lecturers, I gave a full quarter of the presentations. I had assumed that many participants would skip Friday afternoon, so I kept the last lecture for myself, and the session before that was taken by a friend who did not mind speaking to an empty room. But to our delight, the room was full until the very end. More surprising was that everybody attended every lecture. The mathematicians were amazed that what they considered to be safe esoterica was in fact part and parcel of nature. The physicists were amazed that many complicated problems could be solved in a simple and transparent way.

All the Kepler moments of my life to that day had come together.

Fractals Meetings and Birthday Celebrations

A multitude of fractals meetings followed this one. Each became increasingly specialized. I expected this to happen, as did other scientists. I recall a meeting in Trieste where a journalist interviewed me with the meeting's host, Stig Lundquist, then head of the Nobel Committee for Physics. The journalist was astonished to hear that the success of fractals depended on people being familiar with the basic ideas and pushing them in different directions with more specialized topics, resulting in fewer general fractals meetings.

Several of these meetings doubled as birthday celebrations. In 1989, for my sixty-fifth birthday, my physicist friends Amnon Aharony and Jens Feder, with support from IBM France, organized a marvelous meeting, the Fractals in Physics conference. It was held at Mas d'Artigny in Saint-Paul de Vence, high in the hills above the Riviera.

After the meeting, Aliette and I stayed on for a day to allow profound exhilaration to cool off and then, still dizzy, took a short vacation. We drove past the place nearby where I had been a horse groom in 1944, and to fulfill an old curiosity, we splurged on a fabulous dinner at Frères Troisgros, the famed four-star restaurant in Roanne. Then we drove on to Tulle, that hollow in the mountains where I had spent several years during the war, which, after all those years, I still consider my true home.

A bit later, Heinz-Otto Peitgen organized a meeting in Bad Neuenahr. The usual evening talk was replaced by an unexpected treat. My friend, composer György Ligeti, described the deep structure of a piece he had just written. It was part of the series of late piano suites that he did not manage to complete but that became one of the greatest contributions to his repertoire. The score was projected on the screen; his pianist had also been invited and helped the master musician deconstruct, then reconstruct, this very short but unforgettable piece.

On my seventieth birthday, my former postdoc, frequent coauthor, and friend Carl C. Evertsz organized a meeting on the island of Curaçao. His family's prominence there greatly helped, as did the lure of February in the Caribbean. While this meeting was memorable, many speakers came just before their talk and left just afterward. I concluded that the days of the truly interdisciplinary fractals meetings were over. When I was nearing eighty and another meeting started being discussed, I strongly urged my friends not to all meet together but instead to have specialized sessions for each discipline. Some friends heeded my request, and a conference on finance was held at

the Deutsche Bundesbank in Frankfurt for my eightieth birthday. Others did as they pleased and organized an interdisciplinary international meeting in Paris, also for my eightieth birthday. In each case we had a marvelous time.

Riding the Coattails of a Best Seller from Bremen

Essential "promotional" help for the 1982 book came from an exhibit at the University of Bremen and a mass-market book. They revived a tradition of high-class expository books that had long lapsed but used to be practiced by the likes of the great Henri Poincaré.

During the summer of 1984, I was making arrangements to replace our black-and-white graphics with color. Everything was in place when a popular German magazine published an article by Heinz-Otto Peitgen and his colleagues from Bremen featuring precisely the kind of color pictures I was about to undertake. Throughout my life, it had been my principle never to compete frontally with anybody. Therefore, I stopped my work on color and instead wrote to congratulate the authors and suggest that we get together. An exchange of letters followed, culminating with an invitation in the spring of 1985. The Bremen group was preparing a big exhibit of fractal art, to be shown first at home and then to travel around the world. They wanted me to visit Bremen for the vernissage and a lecture. I was delighted to accept. The exhibit catalog was magnificent, became wildly popular, and soon sold out. A foretaste made the cover of *Scientific American*. It was expanded into an extraordinarily beautiful book by Peitgen and Peter Richter titled *The Beauty of Fractals*, for which I was flattered to be asked to write a historical chapter.

I became close to the Bremen group and took part in many of their activities. Some were of a kind I would have hesitated to initiate myself but was happy to participate in. They wrote several textbooks that continue to be basic to the teaching of fractals and chaos. They also organized—both in Germany and in Broward County, Florida—a forward-looking program that uses fractals to help teach mathematics in high schools.

I Become Known as the Father of Fractal Geometry

Let me first mention an overflow that occasioned a shower of papers. U.S. publishers believe that thin books are more attractive. Therefore, *The Fractal Geometry of Nature* was made thinner by printing on high-quality ("Bible") paper and keeping the length under five hundred pages. I was left with a mountain of "cuttings." Papers mentioned in *The Fractal Geometry of Nature* remained to be finished and published. They propelled my publications from a low rate while I was concentrating on *The Fractal Geometry of Nature* to a high rate that lasted several years . . . and has not yet been exhausted.

The Fractal Geometry of Nature generated a formidable wave of interest. The word "formidable" has several meanings—often implying something either promising or threatening—and all those were strongly felt. Invitations of every kind started coming and—amazingly—continue to come. Only a handful could be accepted, but every aspect of my life changed in one way or another. Fads come and go and include best-selling books that soon vanish from shelves and minds. New styles begin slowly but are long-lived.

One reason *The Fractal Geometry of Nature* took off was that an amazing variety of journals reviewed it—in glowing terms. Every time I stopped by the library at IBM Research, or so it seemed, one of our librarians would hand me a new journal, often in a field that I did not expect would know or care about my work. Most unexpected, as I try to think of it, was a periodical put out by the French Royalist Party. Its review began by saying that they found themselves surprised to feel that my book had to be reviewed.

And the book did not become that nightmare of publishers: one that reviewers love but readers avoid. For years, friends who visited bookstores more than I do commented that the science section displayed a few scattered works and a big pile of *The Fractal Geometry of Nature*. It paid for my sons' college tuitions and is still in print.

A Shower of Awards

Are awards important? Having sat on a number of committees, I know all too well that their decisions are not divinely inspired but disconcertingly human. For colleagues pursuing a normal career, awards are one of many other indicators of their progress. Those other indicators being absent in my case, awards took on an altogether different importance, especially those coming as a surprise.

The first two, to IBM's credit, came from inside: an Outstanding Innovation Award at the research division level in 1983 and at the corporate level the next year. Being named an IBM Fellow in 1974 might also be viewed as an early award.

My first outside award was the 1985 Barnard Medal for Meritorious Service to Science. It used to be granted every fifth year by Columbia University, in memory of its longtime president, Frederick Barnard, on the recommendation of a committee of the National Academy of Sciences. Earlier laureates included the likes of Albert Einstein, Niels Bohr, and Enrico Fermi. The previous laureate had been the founder of Bourbaki, my nemesis André Weil! When Ralph Gomory, my manager at IBM, called to announce this forthcoming event, he first asked whether I was sitting down, then read the list of my predecessors. He added that winning this award guaranteed that it would not be my last. Indeed, it was not, and in 1986 I received the Franklin Medal for Signal and Eminent Service in Science, followed by the 1989 Harvey Prize in Israel and the 1994 Honda Prize in Japan.

The Steinmetz Medal, awarded in 1988, was heartwarming because Charles Proteus Steinmetz had been a special hero of Father. Crippled by polio, he rose to be a great inventor and also—as a German liberal who had fled the Second Reich—a great civic reformer.

Particularly exotic was the Science for Art Prize, also in 1988. It was awarded by LVMH Moët Hennessy Louis Vuitton. I was entitled to wonder: Is a provider of booze, even a high-class one, sufficiently respectable, especially given that IBM was still dry at that time? So I responded that I must take a night to consult with my wife. The check was small, but a whole week of festivities in Paris and the provinces was arranged as a public relations effort by a skilled purveyor of luxury. For us, it was unforgettable.

Another not purely scientific award, the Médaille de Vermeil de la Ville de Paris, was supposed to be presented in grand ceremony by the mayor of Paris, Jacques Chirac. But in 1995 Chirac was campaigning to become president of France. As a result, appointments were postponed repeatedly, and finally the medal was handed to me by his successor in one of the grand halls of the Hôtel de Ville. I was asked to prepare both a formal speech for the mayor and my own response. I had passed by the Hôtel de Ville millions of times, and although I should have, I never did suspect that its grand halls were extreme examples of the flamboyant academic style that the Impressionist painters famously rebelled against.

A specialized award in mathematics is the annual Sierpiński Medal of the University of Warsaw and the Polish Mathematical Society. Being selected for this award in 2005 and accepting it in Warsaw marked my belated "closure" with Poland.

The most prestigious award was the Japan Prize for Science and Technology of Complexity, which I received in 2003.

It was a full week of varied and entertaining events, a most enlivening glimpse of cultural Japan. One funny moment: at the gala awards dinner, I was given a translator so I could speak to Her Majesty

the Empress of Japan, who sat next to me. The translator, who stayed kneeling down behind the table for the entire meal, had a very easy job. As it turned out, Her Majesty and I were both fluent in English, and we had a lovely conversation on our own.

★ ★ ★

My most memorable award was the 1993 Wolf Prize for physics, which apparently was triggered by the 1989 Saint-Paul de Vence conference Fractals in Physics. It was memorable for two reasons. First, it was presented to me by Ezer Weizman in his first public function after he became president of Israel. Second, although IBM was ecstatic about the award, I received it at the precise moment in 1993 when pure research was being dismantled.

Full disclosure forces me to report that at Yale, where I would become Sterling Professor, this prize did not impress my physics colleagues. They kept absolutely mum.

Awards Accompanied by Backlash

Sudden success is almost always problematic. The success of *The Fractal Geometry of Nature* failed to make my fledgling discipline intellectually, financially, and organizationally strong. However, it sufficed to make it potentially threatening. The physicist Hans Bethe welcomed an unfair advantage in his scientific work; in my case, it was a keen eye.

But unfair competition from an outsider is something that no group faces rationally. So the worst outcome for *The Fractal Geometry of Nature* would have been that it failed to be noticed. The second worst would have been universal dislike. The third worst, which is what happened, was an uncanny split I had to learn to live with.

On the one hand, it made me a world-renowned scientist, and not by moonlighting as a media personality. Apparently, my solo scientist's work has features that are widely attractive.

On the other hand, I have continually faced strong hostility and criticism. In addition to the continuing flow of glowing reviews there was a trickle of dismissive comments and virulent diatribes.

The Balzac-Bohr-Bialik Syndrome: The Tongue, the Pen, and the Eye

Being an agile writer can be a great asset. Mozart could compose a full opera in his head and know it by heart before sitting down to write it. An opposite extreme case is that of the great writer Honoré de Balzac. He became infamous among typesetters for his peculiar hot-type anticipation of word processing. Having penned a few pages of incomprehensible scribbles with corrections all over, he would send them by messenger to the printer and expect to receive the next morning a galley of what he had written the previous day. To that he proceeded to add further corrections and "bubbles" in the margins, leaving almost nothing untouched, and the process continued several times. Rumor has it that printers assigned to his jobs set up an early trade union to escape spending more than a certain number of hours a day on his demanding work. Once, having seen in a museum in Paris a page of Balzac's proofs—and feeling bubbly and flush—I tried to buy a corrected proof for myself. No such luck, not because—as I feared—they cost too much, but because I could not locate a supplier. The many elite dealers in old books and manuscripts that I consulted didn't know what I was talking about: "Sorry, can't help. Perhaps you should follow the auctions."

The great physicist Niels Bohr is reputed to have been almost as bad—with the added problem that being both wealthy and powerful, he was not much in a hurry. He had to be urged by colleagues to stop revising and publish, and his earlier drafts continue to be viewed as

better than the last and to circulate in a kind of samizdat. Yet another sufferer was the Russian poet Bialik, so this extreme style of writing might perhaps be called the Balzac-Bohr-Bialik syndrome.

I suffer from that syndrome in an acute form. I never begin with a table of contents and then write chapters, sections, and sentences in the order in which they appear. Instead, I start with several already available pieces that can be counted upon to provide the structure of the whole, and I keep adding here and there. Every so often, I wake up in the morning with the overwhelming feeling that a chunk of the book is in the wrong place and had better be brought forward or back. Quite literally, a book does not approach completion until I know it by heart. As a young man, I had no access to a typist or time for careful successive handwritten texts. So I often sent the printer something that in truth was an immature early draft. Galleys required extensive Balzacian changes and sometimes made it preferable for my text to be reset from scratch—leading to very stiff bills. Word processing has made this syndrome incomparably easier to live with but has not cured it. One day, my programmer, watching my assistant suffer with an especially messy draft, asked how I had managed before electronics and without an assistant. My answer: "Extremely painfully."

Let me elaborate by expanding on the distinction I see between "seers," who favor pictures—as I do—and "hearers," who favor language. Written or printed material is a hybrid that came late in human evolution and some otherwise advanced cultures never produced it at all. Hearers like Mozart and Homer put to paper one or several linear sounds heard in the mind's ear—without need for much iteration. I think that Balzac must have been a seer, having simultaneous multidimensional thoughts that demanded being linear during the writing process. My handwriting is poor, and I wonder whether dictation would have helped. The result is seen best after it is set in type. It then becomes subjected to a mental process that resembles metallurgical annealing, where metal dissolves after being kept under fixed conditions. In metaphorical annealing, I find that type reveals unsuspected relationships between words, phrases, paragraphs, or chapters. Once I can see these, I am able to adjust them as needed. The paper becomes a new crucible for creativity, a crutch for lesser Mozarts.

Many scientific articles are completely flat because they are written for people who do not have to be convinced. Their authors are part of a small circle within a well-established domain; they know everybody, or are introduced by their thesis supervisors or mentors, and they write for one another. As a result, style is secondary and unimportant for them. In my case, the fact that I write for an unknown public influences and shapes my style. Whether it is opera or Greek drama, one must know how to enter into a subject quickly because one cannot assume that the audience will wait to understand. One has to be able to speak to people in their style, to motivate and even amuse the reader a little.

This syndrome has caused my scientific productivity to be overly dependent on circumstances that made a helper available. Gaps in my productivity resulted not from a lack of imagination but from a lack of assistance. And I must confess harboring a sharp regret. Had I been able to get more assistance in the early years, I would have moved faster, and *The Fractal Geometry of Nature* would have appeared when money for scientific research was flowing, well before 1982. This would have made a big difference.

27

At Yale: Rising to the University's Highest Rank, Sterling Professorship, 1987–2004

THE ART OF RECEIVING new offers and fast promotions has always baffled me, but I have been lucky on a few occasions. One in particular was landing a job at Yale University.

Adjunct Professor of Mathematical Sciences at Yale

The indispensable intermediary who started the process that led me to Yale was a self-described "institutional economist," Martin Shubik. We had met while I was John von Neumann's postdoc at the Institute for Advanced Study and he was at Princeton University with Oskar Morgenstern (1902–77), Johnny's coauthor on the book *Theory of Games and Economic Behavior.* For a short period in the sixties, we were colleagues at IBM Research, but he soon left for Yale.

Out of the blue, Shubik called me when I was in transit to Harvard in 1964 and again early in 1967. The first time, I must have rebuffed him. The second, I must have sounded more open. Shortly afterward, a call came from the mathematics chairman, Ronald Raphael "Raphy" Coifman. "We know that you have a position at Harvard but keep strong links at IBM and a house in Scarsdale. Both places are far closer to Yale. Could we convince you to join us?" "But what about Serge Lang?" I asked. Lang (1927–2005) was a distinguished mathematician, widely feared for his strong and strongly expressed opinions. "Yes, Serge does have clear opinions of departmental colleagues but keeps them to himself. If you are concerned about what he thinks, this department is the best place to be." "But at this point, I know no mathematician at Yale." "Actually, you met Peter Jones in Stockholm, at Mittag-Leffler." "But he is in Chicago." "Not anymore, he has now moved to Yale. We have not met, but I know your work very well. Plus, Shubik, several economists you know, and other colleagues are working hard to bring you here. Come over. Let's meet and talk."

I went, saw, and was won over. Key attractions were Yale's proximity to our house in Scarsdale and that this would be part of a long-term project. The Yale mathematics department disliked being ranked below Princeton and Harvard, and they had decided to replace "lesser" with "different"—in particular, by expanding less abstract topics. The idea was to first appoint senior people with high name recognition.

The dean of (undergraduate) Yale College was Sidney Altman, a noted biochemist who would soon receive the Nobel Prize. Funds had been collected in memory of Abraham Robinson (1918–74). I happened to have met him, so I knew that he had a good reputation in three distinct fields: aeronautics, symbolic logic, and mathematics. At the time of his death, he was Sterling Professor of Mathematics and Philosophy. Therefore, the kitty was constrained to prospects of exceptional versatility, and I qualified. There was not enough money for a full-time permanent chair but enough for aging me to become— on half-time—the first (and so far only) Abraham Robinson Adjunct Professor of Mathematical Sciences. "Adjunct" contradicts holding a chair, but nobody cared. The negotiations went smoothly because I was helped by experience drawn from my Harvard adventure and was not thinking beyond the original five-year contract. As it turned out, I stayed for seventeen.

Living in New Haven was considered but found impractical because I was mainly at IBM and, later, because IBM continued some perks. It was during these years split between IBM and Yale that IBM Research let go half of its staff and I retired. IBM granted me the title of Fellow Emeritus, the continuing use of my Yorktown office, and a few other benefits that were supposed to last two years but went on for thirteen. So we never moved, and missed much of Yale collegiality—a clear loss.

Commuting by car was tedious, and Aliette went beyond the call of duty by being the driver and enjoying Yale while I was working. The only good story related to commuting is when the architect Philip Johnson invited me to visit him at his famous Glass House. Over coffee, I looked at the rolling estate and observed, "A Connecticut forest is far thicker than that. This looks like the fractal view of Italy as painted by Claude Lorrain. Did you arrange it to fit?" "Of course I did—just look behind you." I turned and saw—unframed and on an easel—what seemed at first glance to be a genuine Lorrain. Dumbfounded, I forgot to ask whether the priceless painting was permanently placed there. Now that Johnson has died and his estate is a museum, I wonder how they deal with such a masterpiece.

Michael Frame, Friend and Colleague

A special pleasure of my Yale years has been the company of Michael Frame. I met him on a visit to his previous place of work, Union College. Soon I invited him to Yale for a year, during which he gave an immensely popular undergraduate course on fractals. After several further visits, he was made an (indispensable) adjunct professor at Yale.

Michael supervises the mathematics introductory courses but also teaches an elementary and an intermediate course on fractals, for which he has prepared an extensive set of course notes. Moreover, he ran very important summer programs for high school teachers. All his courses are extremely popular, we wrote papers together, and our discussions on mathematics and everything else have been one of the nicest aspects of my time in New Haven.

In 2002, we collaborated on the book *Fractals, Graphics, and Mathematics Education,* a compilation of articles written by teachers of fractal geometry. These teachers first gathered in December 1997 at a meeting Michael and I held at Yale. As far as I know, this was the first scientific meeting totally dedicated to the teaching of fractals.

Sterling Professor of Mathematical Sciences at Yale

After twelve years as an adjunct professor, I was given tenure as Sterling Professor. An academic's dream—not only in the United States—is tenure at a great university. But having left the École Normale in 1945, I forgot about academia and moved on. Eventually, I did achieve this dream, but only in the nick of time—in 1999, when I was seventy-five—and on a half-time basis.

"Sterling" is a word with many connotations. It came to matter to me that in the 1920s a grateful Yale alumnus with that family name gave a fortune to Yale. Enough for two buildings to be named after him—a first-rate library and a suitable law school—and professorships chosen by a process whose outcomes ranged from obvious to mysterious. When American academia began to appoint University Professors, Yale merely decided that this ill-defined but always exalted role should be assigned to its Sterling Professors. By a rule inherited from Cambridge or Oxford, Yale must recruit its tenured faculty exclusively from among its own graduates—a requirement harder to amend than to satisfy by a loophole: a special-purpose master of arts degree.

How did this honor affect me? I had often demonstrated the capacity to formulate big dreams that everyone else held to be odd and unreachable—but that I managed to fulfill. A Sterling Professorship of Mathematical Sciences was beyond any such dream, but I was glad to enjoy it as a fitting end to a "march up Mount Parnassus" from such a colorful and crooked path.

Did I perceive the grant of a Sterling as the instant when a maverick cocoon molts into an establishment butterfly? Frankly, I did not. Perhaps because of the studied informality of the event: the president of Yale telephoned, campus mail brought a computer-printed certificate, and the departmental tea served champagne. That was it. Nothing significant changed. I may add that the alumni magazine had planned to feature my arrival in 1987 but did not rush: in fact, it waited long enough to feature together the still-recent Sterling and my forthcoming retirement. The absence of a discontinuity had a deeper reason: I was already at Yale and had no declared enemy.

So Yale was a rousing success where Harvard was not. How to account for this? In terms of awards and membership in academies, the difference between those two mathematics departments was small, but I encountered an altogether different mood. In the 1930s, the Yale mathematics department had been driven by a bitter split between two leading figures: a Norwegian and a Swede, brothers-in-law who became bitter enemies and pushed everyone to choose a side. That dark era was and remains a spur to a strong collegiality.

Isaac Newton Institute

The Isaac Newton Institute for Mathematical Sciences in Cambridge, England, bears some similarities to the Mittag-Leffler Institute in Sweden, but it is larger and of a broader scope. In 1999, from January to April, it held a program on fractals.

The University of Cambridge kindly offered me a visiting Rothschild Professorship, but had to withdraw the offer after finding that I exceeded its retirement age by ten years. However, Gonville and Caius (pronounced "keys") College made me G. C. Steward Visiting Fellow. Quite an experience! Also, the Cavendish Laboratory made me Scott Lecturer in physics. Between those lectures and many semi-

nars, I had my hands full. The Caius fellowship included a furnished house. I had not biked since Tulle and did not dare try to revive that skill, so I took long walks between the house, the Newton Institute, and Caius College; my doctor was pleased.

When I was ready to leave, I was informally told that a long-term visiting fellowship in another college was mine if I was interested. Aliette and I were both extremely tempted, but Yale came up with the Sterling Professorship, which I had no question of turning down, and later grandchildren came and brought us to the other Cambridge, in Massachusetts.

28

Has My Work Founded the First-Ever Broad Theory of Roughness?

HOW CAN IT BE that the same technique applies to the Internet, the weather, and the stock market? Why, without particularly trying, am I touching so many different aspects of so many different things?

An important turn in my life occurred when I realized that something I had long been stating in footnotes should be put on the marquee. I had engaged myself, without realizing it, in undertaking a theory of roughness. Think of color, pitch, heaviness, and hotness. Each is the topic of a branch of physics. Chemistry is filled with acids, sugars, and alcohols; all are concepts derived from sensory perceptions. Roughness is just as important as all those other raw sensations, but was not studied for its own sake.

I started almost from scratch and had to create a new toolbox specifically geared toward the study of forms of roughness that possess certain geometric scaling invariances. Each invariance intrinsically introduces one or more numerical invariants. I reinterpreted one as the first of many quantitative measurements of roughness.

Later, many additional intrinsic measurements were also brought up by fractal and multifractal geometry: it even made a set's "degree of emptiness" into a concrete and useful notion.

In 1982, a metallurgist named Dann Passoja approached me with his impression that fractal dimension might provide at long last a measure of the roughness of such things as fractures in metals. Experiments confirmed this hunch, and we wrote a paper for *Nature* in 1984. It brought a big following and actually created a field concerned with the measurement of roughness. I have since moved the contents of that paper to page 1 of every description of my life's work.

Before my work on roughness, it was either undefined or measured by too many irrelevant quantities. Now it can be measured by one, two, or a few numbers.

The Brownian Coastline Leads Me to the Number 4/3

Weight has long been measured by number, and sensations like color and pitch have long acquired purified forms to which one can attach a well-defined and measurable frequency. But what about roughness? When the great philosopher Plato wrote about sensations, he covered roughness in a mere few lines.

Shortly before I was born, mathematician Felix Hausdorff (1868–1942) assigned to those irregular mathematical shapes called monsters a number he chose to call a "dimension," a word I have referred to in this book. Having heard of its curious mathematical properties, I wondered whether it was irremediably theoretical or if it could be removed from pure esoterica and reinterpreted into something intuitive, concrete, and even practical. It can indeed!

And the icing on this cake is the story of the Brownian island coastline. Brownian motion's ups and downs first shined in Bachelier's ill-inspired but profound model of price variation. Now forget about prices and imagine a point that moves on a piece of paper in such a way that its projections on the left and bottom sides of the sheet are Brownian motions independent of one another. That point is said to perform plane Brownian motion. This concept became widely used in physics and mathematics. But oddly enough, it seemed that no one I ever heard of had examined it on actual samples. So I drew a very long Brownian sample and set myself a challenge. I tried to blend two properties already established by mathematical reasoning and search for new properties that a skilled visual inspection might allow me to observe.

My first efforts were fruitless. Any possible novelty was overwhelmed by a multitude of messy old structures that were begging to be removed.

In particular, my Brownian intervals were deficient—unnatural— from a certain aesthetic viewpoint, one that is familiar in a far simpler shape, as a straight interval's end positions differ from its middle portion. My first step was to eliminate this complication by creating a loop—as when the straight interval from 0 to 1 is made into the circumference of a circle. Obliging the Brownian motion to end where it had started yielded a distinctive new shape I called a Brownian cluster. But it still had too much irrelevant detail, and the elimination of this extraneous complexity demanded one more step. After many false starts, I separated the picture into two parts by "painting" white for all the points in the plane one can reach from far away, and black for those points "screened" from far away by one or another piece of Brownian motion. The result was astounding.

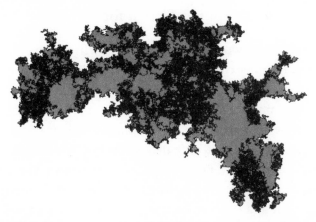

Instantly—but not a second before!—an interesting new island emerged. Automatically, my visual memory recognized some actual islands as well as some islands produced by fractal computer models I had previously devised. The new island's ragged coastline suggested a new concept, that of the boundary of Brownian motion. In the antivisual world of yesterday, this concept had not occurred to anyone. The picture did not "visualize" any existing question. In this case, the picture had to come first and the question later, as a "caption."

Extracting a Brownian island's boundary from Brownian motion achieved something, but less than did the next step, which quantified that resemblance by injecting a numerical measure of roughness. (Contrary to algebraists, who loathe pictures, true geometers accept numbers; we are open-minded to a fault!)

Visually examining the Brownian island's coastline led me to con-

jecture that its fractal dimension is 4/3. We promptly measured it and
got closer to 4/3 than expected. At that point, I conjectured the value
of 4/3 to be mathematically exact.

This experiment was successful in two ways. It confirmed that,
even in a hard science, the eye can be retrained to discern new conjectures that might have escaped algebraic analysis forever. It also gave
mathematics a new direction to follow. Today, the boundary of
Brownian motion might be billed as a natural concept. But this concept could not have progressed to the fractal dimension 4/3 without a
careful visual inspection.

My 4/3 Conjecture Spurs a Search for an Elusive Truth

At IBM, where I was working at the time, my friends went on from
the Brownian to other clusters. They began with the critical percola-
tion cluster, which is a famous mathematical structure of great interest in statistical physics. For it, an intrinsic complication is that the
boundary can be defined in two distinct ways, yielding 4/3, again, and
7/4. Both values were first obtained numerically but by now have
been proved theoretically, not by isolated arguments serving no real
purpose but in a way that has been found quite useful elsewhere. As
this has continued, an enormous range of geometric shapes, so far
discussed physically but not rigorously, became attractive in pure
mathematics, and the proofs were found to be both very difficult and
very interesting.

To prove a purely mathematical conjecture, no number of pictures
or examples suffices, but the value 4/3 is so simple that a rigorous
proof seemed easy—and indeed wonderfully skilled friends promised
to prove the conjecture overnight. They revised this promise to next
week, next month, next year—and finally to the twenty-first century
and third millennium.

The proof turned out to be extraordinarily tricky, and the success
of an eighteen-year worldwide search created an enormous sensation
and generated great enthusiasm and activity. The three mathematicians who combined their skills to achieve it won instant acclaim, and
in 2006 the youngest of the trio received the prestigious Fields Medal,

the award for exceptional mathematical promise. Not only did the difficult proof create its own very active subfield of mathematics, but it affected other, far removed subfields by suddenly settling many seemingly unrelated conjectures.

It was the first Fields in probability theory, but in previous years my key conjecture concerning the Mandelbrot set had already led to two Fields. In an ironic way, my disregard of the customary division of labor has advertised that, in mathematics, the labors of conjecturing and proving may gain by being divided.

My Work Reaches a New Audience

All of this activity has taken me around the world, lecturing, meeting with groups, and showing my pictures.

I often hear comments like this from people of all ages: "May I shake your hand? In this country, your fractal geometry is discussed in high school. So we all first heard your name and saw your mathematical pictures several years ago, and we just assumed—without thinking—that you have long been dead. You might have lived shortly after Newton. We can't believe that we could actually hear you discuss how part of our schoolwork had first come to your mind. To shake your hand would be a strange experience . . . a big event."

Words from a charming young lady seemingly representing a group of college students who had packed a lecture I had just given. Of course, I was glad to shake that young lady's hand. Uncanny forms of flattery! Each lifted me to seventh heaven! Truly and deeply, each marked a very sweet day! Let me put it more strongly: occasions like that make my life.

29

Beauty and Roughness: Full Circle

A MEMOIR IS A LESSON IN HUMILITY. I was born in 1924, and it is now 2010. To put my personal achievements in perspective, those dates matter indeed. The Great Depression dominated the earliest world news that I recall, and another depression threatens to dominate my last days. My late adolescence coincided with World War II, which I spent in the impoverished hills of central France. My survival was continually threatened, but my dreams ran free and seeded my future.

Does it matter that I stumbled into IBM Research when its golden age began and stayed until the day it ended? That it is where my wartime dreams finally managed to be realized?

Of those born in the year 1924, I am sure that many became scientists. What made me seek out a role that others missed or spurned? I have always wondered, and I wrote this book in an effort to understand myself.

When I turned thirty-five, I questioned my life. Had I, in my dreams of leaving my mark on science, really "missed the boat"? I am keenly aware that this fear led me to reinvent myself surprisingly late in life, when I did my best-known work. My refoundation of finance was to occur as I neared forty, and the discovery of the Mandelbrot set came at fifty-five. For a scientist, those are unusually—astonishingly—old ages, as many witnesses have noted. And the number of would-be role models I have considered but not followed has been heartbreakingly large.

Had my work on price variation been accepted in the 1960s, I might have settled as a satisfied slave of my creation. Who knows. But events proceeded differently. I was expelled to resume my wandering intellectual life. No official Galilean trial had to punish me for attempting to propagate disallowed ideas. No one was listening, and I had no need to turn my face away to say, very softly, of Earth, "Still it turns."

What has attracted me to problems that science either had never touched or had long left aside—continually making me feel like a fossil? Perhaps a deficit in regular formal education. My adolescence during the wartime occupation of France was illuminated by obsolete books, ancient problems long abandoned without solution, and timeless interrogations. The form of geometry I increasingly favored is the oldest, most concrete, and most inclusive, specifically empowered by the eye and helped by the hand and, today, also by the computer. In due time, it turned out to be an elusive point where formula and picture meet on even terms, where theory meets the real world, and where mathematics and hard science meet art so that their worth and beauty shine far beyond the narrow world of experts, bringing an element of unity to the worlds of knowing and feeling.

Since I became a scientist, much of my work has consisted of bringing a medley of old issues back to life and triumphant evolution. While they seemed to share little beyond common antiquity, they all eventually revealed themselves as being concerned with roughness in nature and art. Surprisingly, a loop seems to have been established between structures that were first identified for mere decoration, then, much later, introduced by mathematicians for the purpose of pathology. Again, even later, these same structures were used by me for the purpose of science and, unwittingly, as a bonus, for the purpose of creating beauty.

Roughness in Painting and Music

My earliest fractal "forgeries" of trees and mountains made me wonder: If nature's real trees and mountains are indeed fractal, should not the same be true of their representations by painters? Think of Leonardo da Vinci's celebrated drawing *A Deluge*, reproduced in *The Fractal Geometry of Nature*. Unquestionably, it is fractal.

Skilled artists must find arrangements, like mixtures of eddies of all sizes, that look balanced; does not that mean that elements of all sizes are distributed in a natural—that is, fractal—way?

The Fractal Geometry of Nature reproduced Hokusai's print *The Great Wave,* the famous picture with Mount Fuji in the background. Hokusai was at his peak around 1800, but history provides examples of many earlier painters or philosophers who were aware of complicated shapes with fractal structure. Claude Lorrain, a French painter who worked mostly in Italy, painted landscapes that claim to be realistic, but in fact are extraordinarily simplified and easily interpreted in fractal terms. Historically, painters have always seen the possibilities of fractal structure, but it did not develop into a geometry since very few wrote about it and probably none read about it.

The Russian painter Wassily Kandinsky (1866–1944) was filmed as he worked on a sheet of paper about three feet square. He began with a slash across the whole and then added shorter slashes. When the film stopped, he was at work on many even shorter slashes, confirming a feeling I had looking at Kandinsky's paintings: he understood fractality—perhaps not explicitly, but intuitively.

Initially, I viewed these works of art as amusing, though not essential. But I soon changed my mind as innumerable readers made me

aware of something strange. I began to recognize fractals in the works of artists since time immemorial. A remarkably large number of artists had no vocabulary to express their grasp of the nature of fractals, yet such understanding comes through clearly in their work.

One mathematical structure I called the Sierpiński gasket, made of several identical parts, turns out to be very common in decoration in Italian churches, either in mosaics on pavement or in paintings on the roof and ceiling. Other fractal structures were found in Persian and Indian art of different periods.

I have strong connections with composers, who inhabit an entirely different world. In particular, György Ligeti confided to me that until he saw my pictures, he had not understood an important aspect of music: it is not free to do as it pleases, because it must be fractal. The schools of music never taught how to distinguish music from noise. When Ligeti received a prize in New York, a major article appeared in which he listed the greatest designs ever. The list included the Book of Kells and the Taj Mahal . . . and the Mandelbrot set! That was an extremely strong statement, and I was pleased to meet him shortly afterward. We have had interesting times together, including serious public discussions. Charles Wuorinen is another widely known contemporary composer who understands fractality. He liked to say he had used a fractal approach to composition for some time. He was well aware that much of Western music exhibits similar structures over different time scales. Wuorinen and I did an extraordinary show at the Guggenheim in 1990 called Music and Fractals. It is fascinating to see how two people from such different cultures can collaborate, if they have the desire to do so.

Unrelated Deeds or a Unified Fractal Approach to Roughness?

Isaiah Berlin (1909–97), a British philosopher and man of action— whom I met—has written about the distinction the ancient Greek writer Archilochus drew between the fox, who knows many things, and the hedgehog, who knows one big thing. Once, colleagues assigned to introduce me before a lecture kept asking whether I viewed myself as a fox or a hedgehog. The point is that they all saw me as having two faces.

The marble sculpture, below, represents Janus, the Roman god of doorways and bridges. He was believed to have two opposite and contrasting faces—one to judge and perhaps to repel, the other to welcome and attract.

For better or worse, two faces are also an appropriate metaphor for one of mankind's greatest accomplishments—and my field of study—

which is broadly interpreted as the mathematical sciences. Its judging face is that of a purist, a specialist taking pride in thoughts that the bulk of humanity views as dry and cold. Its welcoming face is actually something of a blur of many roles that mathematics plays in the labors and pleasures of daily life: architecture, engineering, and the arts. Symbolically, they look simultaneously into the future and the past.

Being myself a faithful—though by reputation a turbulent—servant of mathematics, I have continually rebelled against those two faces looking in opposite directions. I rejoiced in learning that, many centuries ago, the two faces had been turned in the same direction, peace prevailed, and splendid fruit came forth.

Fractal geometry is one of those concepts which at first invites disbelief but on second thought becomes so natural that one wonders why it has only recently been developed.

Real Roughness Is Often Fractal and Can Be Measured

The foremost measure of roughness is fractal dimension. The simplest form of fractal dimension is the similarity dimension, and the earliest illustration of this is a curve provided by Helge von Koch. Because its length is infinite, the Koch curve began as one of those monsters—or toys, as I refer to them. Fractal geometry brought out the wonder by setting it to the task of describing coastlines and then mastering nature. An even more monstrous monster appeared when

296

Giuseppe Peano constructed a "curve" that visits every point of the plane. It created a storm among mathematicians and a deep split between purist extremists and those who care about the real world. A universally held opinion was that the Peano curve was totally nonintuitive and extravagant. These were words not of disappointment but of great pride on the part of pure mathematicians. The illustration below combines the Koch curve (the outline, or coastlines) and the Peano curve (the rivers, or blood networks, across the surface of the plane).

Unimaginable privilege, I participated in a truly rare event: pure thought fleeing from reality was caught, tamed, and teamed with a reality that everyone recognized as familiar. Monsters were made into servants—in the manner that Kepler pioneered when he showed that planets' orbits fitted ellipses, which had been the ancient Greeks' playthings.

Roughness is ubiquitous in nature and culture—found in the distribution of galaxies and in the shapes of coastlines, mountains, clouds, trees, and the various ducts in the lungs; also in stock-price charts, paintings, music, and several mathematical constructions (well-known ones and those I fathered). Less familiar but worth a mention: the roughness of clusters in the physics of disorder, turbulent flows, chaotic dynamical systems, and anomalous diffusions and noises. These are typical of the many topics I studied.

Like smooth shapes exemplified by the ideal circle, mathematical fractals are described by absolutely precise formulas that the computer can implement, as closely as one wishes, with very concrete objects: pictures. Each picture led me to specific insights into a specific area of science and art. Some pictures proved to have a profound and durable impact and were expanded by several very focused investigators.

The set of color illustrations in this book are varied in many ways. Some are natural, some are works of art, but most are purely mathematical constructs drawn by computer with the help of appropriately chosen formulas. Those formulas share an essential property that I spent my whole active life investigating from all sides: roughness. They are not drawn merely to be pretty, but to serve all kinds of purposes in all kinds of sciences. This is why several seem realistic, reminding us of shapes in nature. They are all fractal, which is why, when asked what I do, I call myself a fractalist.

To appreciate the nature of fractals, recall Galileo's splendid 1632 manifesto: [Philosophy] *is written in the language of mathematics, and its characters are triangles, circles, and other geometric figures, without which . . . one is wandering about in a dark labyrinth.* Observe that circles, ellipses, and parabolas are very smooth shapes and that a triangle has a small number of points of irregularity. These shapes were my love when I was a young man, but are very rare in the wild. Galileo was absolutely right to assert that in science those shapes are necessary. But they have turned out not to be sufficient, "merely" because

most of the world is of great roughness and infinite complexity. How-ever, the infinite sea of complexity includes two islands of simplicity: one of Euclidean simplicity and a second of relative simplicity in which roughness is present but is the same at all scales.

The cauliflower is the standard example of shapes that appear more or less the same at all scales. One glance shows that it is made of florets. A single floret, examined after you cut away everything else, looks like a small cauliflower. If you strip that floret of everything except one floret of a floret—very soon you must take out your mag-nifying glass—it is again a cauliflower. A cauliflower shows how an object can be made of many parts, each of which is like a whole, but smaller. Another example of this repeated roughness is the cloud. A cloud is made of billows upon billows upon billows that look like clouds. As you come closer to a cloud, you get not something smooth but irregularities on a smaller scale.

Do I claim that everything that is not smooth is fractal? That frac-tals suffice to solve every problem of science? Not in the least. What I'm asserting, very strongly, is that when some real thing is found to be unsmooth, the next mathematical model to try is fractal or multi-fractal. Since roughness is everywhere, fractals are present every-where. And very often the same techniques apply in areas that, except for geometric structure, seem completely independent.

Fractals Have Been Here Forever and Now Have a Home

For the most part, there was no place where the things I wanted to investigate were of interest to anyone. So I spent much of my life as an outsider, moving from field to field. Now that I look back, I realize with wistful pleasure that on many occasions I was ten, twenty, forty, even fifty years ahead of my time. Until a few years ago, the topics in my Ph.D. dissertation were unfashionable, but they are very popular today.

My ambition was not to create a new field, but I would have wel-comed a permanent group of people with interests close to mine and therefore breaking the disastrous tendency toward increasingly nar-rowly defined fields. Unfortunately, I failed on this essential point. Order doesn't come by itself. In my youth, I was a student at Caltech

while molecular biology was being created by Max Delbrück, so I saw what it means to bring a new field into existence. But my work did not give rise to anything like that. One reason is my personality—I don't seek power or run around asking for favors. A second is circumstances—I was in an industrial laboratory because academia found me unsuitable. Besides, establishing close, organized links between activities that otherwise are very separate might have been beyond any single person's ability.

I did not plan any general theory of roughness, because I prefer to work from the bottom up, not from the top down. So even though I didn't try to create a field, now, long after the fact, I am enjoying this enormous unity and emphasize it in every new publication.

I reach beyond arrogance when I proclaim that fractals had been pictured forever but their true role remained unrecognized and waited to be uncovered by me.

As my wandering life fades away, I keep thinking of the wild ambition to survive and shine that has pushed me since adolescence. Each partial success aroused some old expectation or some old hunger. Ironically, this same pattern is one I have often dealt with in my research. Even at this late stage, I suffer when some event reawakens an old fleeting hope I had to leave untested. In the words of George Bernard Shaw:

The reasonable man adapts himself to the world;
the unreasonable one persists in trying to adapt the world
to himself. Therefore all progress depends on the
unreasonable man.

Only late in life did I see this quote, and—to help knowledge and reason advance—I had been quite unreasonable all along.

* * *

You have now heard my story. Does not the distribution of my personal experiences remind one of the central topic of my scientific work—namely, extreme fractal unevenness? All counted, I have known few minutes of boredom. It has been great fun, and to some extent the fun continues. What else could one have asked for?

To be close to my grandchildren, I have retired from Yale, closed my IBM office near New York, and endured the agony of downsizing

from a big house to a Boston apartment. As I have always known, uprooting can be rational but is never sweet.

An old man by now, past two times forty, I see myself continuing in some ways to mature. And to remain embattled. How come? Perhaps by fluke, but I think mostly for a reason.

Afterword

Michael Frame,
professor of mathematics, Yale University

BENOIT MANDELBROT died shortly before he could make final revisions to this memoir. Aliette, his wife of many years, asked me to write this afterword. I hope I can offer a slightly different perspective on how Benoit's work fits into the worlds of science and culture.

★ ★ ★

I met Benoit twenty years ago when he hired me to join his group at Yale. I think he brought me into his world because, in a specific sense, we both were little kids. Benoit would call with a question or an idea and we'd be off. Then I'd glance at the clock and an hour or two would have passed. We'd go our separate ways for a week or so, working out some details of what we'd seen. He'd call and we'd be off again. I think I shared his sense of innocent wonder.

Benoit loved complicated things, the roughness of coastlines and price graphs, the music of Charles Wuorinen and György Ligeti, the paintings of Augusto Giacometti and prints of Hokusai. What he saw in all of these, what may have helped guide his thinking through the wanderings of his long (but still too short) career, was a sense that there were common features to all these examples. Patterns that kept recurring as he looked ever closer. To be sure, many scientists and artists had noticed this, and the examples of continuous, nowhere differentiable curves were familiar from basic real analysis courses. But Benoit saw much more, a way to quantify these recurring patterns so that complicated shapes might be easily understood dynamically, as processes, not just as objects.

The power of this paradigm is immense, and still persists. In Sep-

303

tember 2010, I had the pleasure of watching the eighty students in my fractal geometry course enter the classroom not knowing how to generate fractal images, follow some simple steps, and by the end of the class period be able to tell me how to generate these fractals just by looking at the images.

I pointed out just how much their understanding had grown that day. Their looks of surprise gave way to grins and "cools." (Then I warned them not to expect such miracles every day.) This is what Benoit gave the mathematical world. If anyone doubts the power of his gift, compare a standard geometry class lesson on plane transformations with this day in a fractals class.

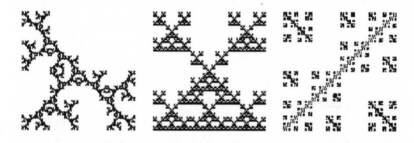

What makes this effective is the visual complexity of the images—reflecting Benoit's inspiration—together with the ability to decode these images into a few simple rules. The other point is that once a student learns how to see these patterns, the solution can be tested in seconds using basic software. Visual experimentation implemented by computers—another of Benoit's initially unpopular causes—is now so commonplace that it warrants no remark.

Outside of science, Benoit probably spent the most time on fractals in finance. Bachelier's 1900 model exhibited three properties: scaling, independent jumps, and short tails (large jumps are rare indeed). The first of these properties fit into, and perhaps helped form, Benoit's view of the fractality of financial series. The other two properties do not agree so well with observations. Benoit's 1960s studies, fractional Brownian motion and Lévy stable processes, also are scaling. The first has dependent jumps but still has short tails; the second has long tails but independent jumps. In the 1990s, Benoit developed an extraordinarily simple and elegant approach, the multifractal cartoon. These cartoons are scaling, have dependent jumps and long tails, and can be fine-tuned easily. Much of Benoit's work was based on a simple idea—scaling, iteration, and dimension—applied with great finesse in new settings.

By far, the biggest surprise is the Mandelbrot set. In class we set up the simple formula and describe the iteration process and how to color-code the result. Then we run the program and wait for the shock to spread across the room. "This formula produces *that* picture??? Are you kidding me?" "Just wait, you haven't seen anything yet. Let's magnify a bit and see what we find." "You mean those complicated twirls and swirls *still* come from the same little formula?" "Absolutely." This miracle needs no further discussion. Look at the pictures, remember the simple formula, think, and be amazed.

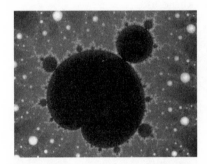 

Benoit wrote and talked about fractal patterns in art, architecture, music, poetry, and literature. He was overjoyed to find fractal aspects—surely discovered thanks to a careful eye for the subtle patterns of nature—in the art of Hokusai and Dalí, the architecture of Eiffel, the music of Ligeti and Wuorinen, the verses of Stevens, the plays of Stoppard. Every new instance of fractals pleased him. Of course, Benoit was clear, often blunt, in his criticism of misapplications of his ideas. Given the range of these mistakes, his ire was understandable.

After Benoit told me of his illness, I tried to get him to reflect on his accomplishments, his amazing legacy—enough to satisfy a dozen brilliant scientists. Instead, he talked about the work that remained to be done, work that would go unfinished, at least by him. What he regretted most, other than not completing this memoir, was leaving his ideas about negative dimensions at such an early stage. As with so much of Benoit's work, this began with a simple but elegant question. Starting with the familiar formula for the dimension of the intersection of two sets, Benoit asked, "Can we make sense of a negative result from the intersection formula?" From simple questions . . . to calculations of clever examples . . . to validation by experimental work.

Another unfinished area, lacunarity, began when Benoit noted that many fractals appearing quite different have the same dimension, like the fractals you see below. Benoit wondered if the distribution of gaps in a fractal could be measured by some number. This proved challenging, but early steps have been taken. More needs to be done. Benoit talked about how the project could continue. In the end, he cared about the work, not about his reputation. I believe this was true throughout the long arc of his remarkable life.

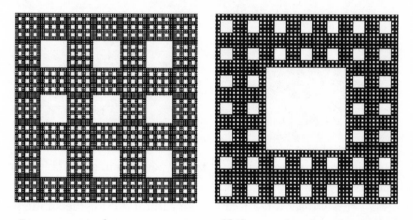

One summer long ago, my grandfather and I were lying on his driveway watching the stars come out as the sky darkened. "Darkened" isn't the right word—it isn't that the evening sky is darker; it's deeper. I was looking into impossible depths. I had a glimpse of an amazing surprise just slightly out of reach, as if I were waking up. What was it? Half a century later, I learned it from Benoit. People often have brief hints of radically different ways to organize what they see, hear, and feel. Very few have more than a glimpse. Benoit shifted the whole world under our feet, giving thousands of people the tools to see the world in a new way. Learning how to recognize this is the clearest example I know of waking up.

Benoit's lesson is this: Find the thing you love and follow it with all your heart. What you are following may not always be clear, but if you persist, you will find it, and when you do, you will wake up. What we find is ours, and what each of us finds enriches all of us. This, I think, is Benoit's last, best lesson. Follow your curiosity, your passion, wherever it leads. Whether you find a new world or a new snowflake, it doesn't so much matter. Like fractals, life is better understood as a process than as a result.

Often Benoit said a fractal is defined as well by what has been removed as it is by what remains. Benoit's dying has left a hole. His wonderful curiosity, sense of kindness, fierce loyalty to friends, and unbounded love for his family have dissolved into the night sky. Memories remain, of course, and a monumental list of accomplishments.

★ ★ ★

The last word is Benoit's, from what was to be his last major talk, at the February 2010 Technology, Entertainment, and Design conference in Long Beach, California:

Bottomless wonders spring from simple rules . . . repeated without end.

Index

Page numbers in *italics* refer to illustrations.

Boeing 707, 120
Bohr, Niels, 124, 272, 276
Boniusiowa (maid), 22
Book of Kells, 295
Boulez, Pierre, 137
Bourbaki cult, 18–19, 89, 92–3, 94, 95, 98, 110, 254, 272
Brahe, Tycho, 156
Brandenberg Concertos (Bach), 136
Brard, Roger, 112–13
Braude, Mrs., 35
Braudel, Fernand, 194
Brazil, 97
Bremen, University of, 270–1
Brooklyn Polytechnic Institute, 174
Brownian motion, 160, 218, 247
    fractals and, 286–8, 304
Brownlee, John, 136
Bruhat, Yvonne Choquet, 143
Bruno Boccanegra de la Boverie (dog), 185
bubble chamber, 123
Buddenbrooks (Mann), 71
Busch, Fritz, 136

Caesar, Julius, 46, 233
Calcomp, 210
calculus, 70
California, University of
    at Los Angeles, 171
    at San Diego, 231
Caltech, 108, 109, 121, 132, 146, 179, 299–300
    aircraft design project of, 119–20
    BM's anonymous benefactor at, 115–16
    BM's choice of, 112–14
    BM's journey to, 114–15
    curricula of, 116–18
    faculty of, 116–19, 123
    intellectual life at, 123–4
    Inter-Nations Association of, 123
    math curriculum of, 116–17
    tuition at, 115–16
Calvin, John, 184
Cambridge University, 283–4
Camp de Cazaux, 130
Camp de Château Bougon, 129

Camp de la Folie, 128–9
Candide (Voltaire), 238
Cantor, Georg, xv
    on mathematics, 178
Carceri (Piranesi), 184
Carleson, Lennart, 247
Carmen (Bizet), 136–7
Carnot, Lazare, 109, 206–7
Carnot, Sadi, 38, 109
Carrier, George, 237
Carvallo, Moïse Emmanuel, 86
Casimir, Hendrik, 147–8
cathode ray tube (CRT), 211
Catholic Church, 97
Cavendish Laboratory, 283
Center for Astrophysics and Space Sciences, 231
Center for Genetic Epistemology, 187
Chaconne for Violin (Bach), 136, 137
Chaitin, Gregory, 261
Chamberet: Recollections from an Ordinary Childhood (Morhange-Beque), 49
chance, theory of, 94
Chansiergues d'Ornano, Suzanne de, 73
chaos theory, 118, 257
Charlier, Carl, 230
Charpentier, Yves, 134–5
Chicago, University of, 121–2, 126, 234, 237, 238, 239, 242–3
    job offer from, 225–6
Chicago Symphony Orchestra, 268
China, 67, 230
Chirac, Jacques, 273
Chomsky, Noam, 164, 240
Chopin, Frédéric, 26
Churchill, Winston, 103, 222
Church of Saint-Martin (Tulle), 51–2
Cicero, 46
Citroën, André, 185
Clermont-Ferrand, University of, 52, 58–9, 80
coastlines, 212, 286–8, 297
Cocke, John, 207
coding theory, 174
Coifman, Ronald Raphael, 279
Coissard, Monsieur, 69, 71
Collège de France, 93–4, 141
    author as lecturer at, 243–4
    job offer from, 244–6

television industry, 147
Teller, Edward, 114, 168
"Test of Mandelbrot's Stable Paretian
    Hypothesis, A" (Fama), 226
Theory of Games and Economic Behavior
    (von Neumann and Morgenstern),
    126, 159, 279
thermodynamics, 109, 118–19, 123, 142,
    164, 171–2
    statistical, 140, 154–5, 165
Thill, Georges, 136
Thomas, Harold, 234
Thomas J. Watson Research Center, 203
    see also IBM
Tintin (comic book), 235
Tisza, László, 164–5
Titian, 37
Tolman, Richard Chase, 118–19
Trans World Airlines, 114
Treasury Department, U.S., 226
Treaty of Riga (1921), 28
Trilling, Leon, 181, 239
Tronchon, Marie-Thérèse, 56–7, 61
Trotsky, Leon, 26
Trumbull Lectures, 243
Tuckerman, Bryant, 208–9
Tulle, France, 50–6, 59, 62, 66, 269
    daily life in, 54–6
    German atrocity in, 80–1
Tunisia, 176
turbulence, 222, 224–5, 234, 237, 247
    Hausdorff dimension and, 240–1
    multifractal study of, 241
Turkey, 194

Ukraine, 5, 25, 28
ultraviolet catastrophe, 154
Ulysses (Joyce), 155
Union Sacrée (horse), 75
United States, 10, 26, 46, 58, 88, 97, 110,
    139, 152

Van Ness, John W., 230
"Variation of Certain Speculative Prices,
    The" (Mandelbrot), 219, 224–5
Vatican, 61

Vatican Museum, 37
Venus de Milo, 71
Vichy France, 49–50, 58, 67
Victory of Samothrace, 71
Vienna Philharmonic, 136
Vietnam, 103, 176
Voltaire, 71, 238
von Neumann, John, 91–2, 114, 126, 159,
    169, 170–1, 174–5, 177, 179, 180,
    188, 205, 215, 254, 279
    background and achievements of,
    168–9
    BM's thesis defended by, 167–8,
    172–3
Voss, Richard, 212, 242, 267

Wahl, Jean, 19
Wallace, Henry, 135
Wallis, James R., 230
Walsh, Joseph L., 151, 154, 156–7
Walter, Bruno, 134
Warsaw, Poland, 25, 37, 41
    BM's childhood in, 21–3
    Great Depression in, 32
    Saxon Gardens of, 22, 40
    World War II and, 9
Warsaw, University of, 273
Warsaw Ghetto, 4, 6
Watanabe, Michael, 205
weather, 222, 254
Weaver, Warren, 170–1
Weil, André, 18, 89, 93, 98, 254, 272
Weisskopf, Victor, 124, 235
Weizman, Ezer, 273, 274
What Is Life? (Delbrück), 124
W. H. Freeman, 266–7
Wiener, Norbert, 95, 126, 159, 165,
    179–80, 218
    Szolem Mandelbrot's relationship
    with, 160–1
Wiesner, Jerome B., 161–2, 164, 238–9
Wigdorczyk, Mrs., 29
Wigner, Eugene, 114, 168, 178
Wilson, Robert, 171
Witte, Sergei Yulyevich, 12
Wolf, Captain, 103–4
Wolf Prize, 181, 274

Illustration Credits

A graduate of the École Polytechnique, Benoit Mandelbrot received his doctorate in mathematics from the University of Paris and spent thirty-five years at IBM as a research scientist and seventeen years as a professor at Yale University. Best known as the father of fractal geometry, he transformed our understanding of phenomena characterized by extreme variability and roughness in a wide array of fields including economics, data visualization, fluid turbulence, biology, geology, and material science. He died on October 14, 2010.

A NOTE ON THE TYPE

This book was set in Monotype Dante, a typeface designed by
Giovanni Mardersteig (1892–1977). Conceived as a private type
for the Officina Bodoni in Verona, Italy, Dante was originally cut
only for hand composition by Charles Malin, the famous Parisian
punch cutter, between 1946 and 1952. Its first use was in an edition of Boccaccio's *Trattatello in laude di Dante* that appeared in
1954. The Monotype Corporation's version of Dante followed in
1957. Although modeled on the Aldine type used for Pietro Car-
dinal Bembo's treatise *De Aetna* in 1495, Dante is a thoroughly
modern interpretation of the venerable face.

Composed by North Market Street Graphics,
Lancaster, Pennsylvania

Printed and bound by Berryville Graphics,
Berryville, Virginia

Designed by Virginia Tan